A SOURCE BOOK OF AIRCRAFT

ACKNOWLEDGEMENTS

The author and publishers are indebted to Mr. John Taylor, Mr. Colin Munro and the following for the use of their illustrations in this book.

Air France—139; Air Photo Supply—63, 82; B.O.A.C.—150; The Boeing Company—36, 56, 64, 71, 89, 92, 104, 127, 133, 136, 149, 157; British Aircraft Corporation—30, 37, 51, 75, 80, 91, 116, 126, 135; British European Airways—122, 156; Camera Press Ltd.—128; De Havilland—38, 110, 111, 125; Dornier—58; Flight International—15, 25, 48, 55, 57, 70, 88, 112; Fokker-VFW—140; Handley Page Ltd.—44, 60, 123, 132; Hawker Siddeley Aviation—34, 40, 42, 49, 59, 61, 67, 99, 103, 107, 115, 118, 119, 124, 130, 131, 146, 147; Imperial War Museum (Crown Copyright)—43, 45, 92; Lockheed Aircraft Corp.—66, 86, 129, 143; Lufthansa—62; McDonnell Douglas—144, 145, 152; North American Aviation—100, 121; Piper Aircraft Co.—76; R.A.F. (Crown Copyright)—41, 134; Ryan Aeronautical Co.—53; Science Museum (Crown Copyright)—16; Short Bros. & Harland Ltd.—32, 52, 96; United Air Lines—73; Westland Aircraft—68.

A SOURCE BOOK OF

AIRCRAFT

COMPILED BY MAURICE ALLWARD

WARD LOCK LIMITED
LONDON AND SYDNEY

ISBN 0 7063 1217 1

Ward Lock Limited, 116 Baker Street,
London WIM 2 BB

MADE IN ENGLAND
PRINTED IN GREAT BRITAIN
BY BUTLER & TANNER LTD
FROME, SOMERSET

INTRODUCTION

This source book covers the story of aircraft from the historic Wright Flyer of 1903 to the supersonic Concorde of today.

The pictures—all types of aircraft from all countries—provide a unique guide to the way aircraft have developed, and clearly show the big impetus to development given by the two world wars.

At the beginning of the First World War aircraft were frail machines, unarmed and suitable only for reconnaissance duties. By its end pilots in swift, heavily armed fighters were shooting each other out of the sky over the Western Front, bombers were showering high-explosive and incendiary bombs on battlefield and town alike; the first dive-bombers had attacked pin-point targets; at sea, torpedo-planes had scored their first victories.

Equally spectacular changes took place during the Second World War. In five years, fighter fire power grew from rifle-calibre machine-guns to batteries of cannon, sometimes of 30-mm. calibre and larger, and the first air-to-air missiles had been fired. In the same period, bomber development led to aircraft capable of dropping 22,000 lb. 'Earthquake'. bombs.

The book also shows how many of the best and most important aircraft produced over the last half century were developed by Britain. This fact should be remembered when people discuss whether or not Britain should have a strong aircraft industry.

Included are many of the aircraft usually considered important to the story of aviation development, such as the Eindekker Scout, DH4, Dakota, Me 109, Spitfire, Lancaster, Mig 15, Boeing 707 and the Harrier. But space has also been found for many lesser known types which although not of great historical significance, nevertheless played their part in the panorama of aircraft development. Such aircraft include Mignet's Flying Flea, a brave but unsuccessful attempt to bring flying within reach of the average "Do-it-yourself" enthusiast, and the Heston Phoenix of 1935 which, with its retractable undercarriage, was one of the smartest light aircraft ever built and which, even today, does not appear out of date.

The book also shows some of the latest types produced in the Soviet Union, indicating the efficiency and strength of the aircraft industry of that country. The Draken and Viggen, from Sweden, show that a small country can compete effectively with the big ones, if it sets its mind to do so.

LANDMARKS IN THE HISTORY OF FLIGHT

1783   First manned balloon ascent made by Pilâtre de Rozier and the Marquis d'Arlandes.

1784   First balloon ascent in Britain made by James Tyler, at Edinburgh.
Jean Blanchard, one of the great pioneer balloonists, and Dr. Jeffries crossed the English Channel.

1802   First parachute descent made by Garnerin (in England) from a balloon over London.

1809   Sir George Cayley experimented with a glider in England.

1842   W. S. Henson designed the first powered aeroplane—the Aerial steam carriage.

1852   Giffard made a flight in an elongated balloon steered by a rudder and propelled by a steam-engine driving a propeller.

1873   Joystick designed.

1896   Otto Lilienthal, known as 'the father of the aeroplane', killed in an accident after a series of gliding experiments.

1900   Zeppelin's first airship made its trial flight.

1903   Orville Wright made the world's first controlled, power-driven aeroplane flight at Kitty Hawk, North Carolina.

1905   First officially recorded flight was made by Wilbur Wright at Daytona, Ohio. He flew 11–12 miles in 18 minutes 9 seconds.

1906   Santos-Dumont, in France, made the first officially recorded aeroplane flight in Europe.

1909   J.T.C. Moore-Brabazon (later Lord Brabazon of Tara) made first officially observed aeroplane flight in the British Isles.
M. Louis Blériot, a Frenchman, flew the Channel from Les Baraques, near Calais, to a point near Dover Castle, a distance of 32 miles.

1910   Zeppelin completed his first passenger airship. E. T. Willows, in his airship 'Willows III', made the first airship flight from England to the Continent.

Santos-Dumont *Demoiselle* ▶

LANDMARKS

1911 First airmail in the United States. Experimental airmail service operated between Hendon and Windsor in United Kingdom.

1918 Handley Page V/1500 biplane flew over London carrying 40 passengers and sufficient fuel and oil for a six-hour flight. Handley Page biplane completed flight from Cairo to Delhi in 57 hours' flying time.

1919 John Alcock and Arthur Whitton Brown (both later knighted) made the first direct Atlantic crossing by air, flying from Newfoundland to Ireland, a distance of 1,890 miles, in 16 hours 12 minutes. British airship R34 made the first east-west air crossing of the Atlantic—from Scotland to Long Island, New York, and back again—carrying a crew of 30. Ross and Keith Smith and two companions, in a Vickers Vimy biplane, made the first aeroplane flight from England to Australia (distance 11,294 miles), in a flying time of 124 hours.

1926 Commander R. E. Byrd, with Lloyd Brunett as pilot, flew from King's Bay, Svalbard (Spitzbergen), to the North Pole and returned in 15 hours.

1926 The Amundsen-Ellsworth-Nobile expedition flew the semi-rigid airship 'Norge' from King's Bay, over the North Pole, to Teller, Alaska, about 2,700 miles.

1927 Colonel Charles A. Lindbergh, United States, made first solo crossing of the Atlantic, flying from New York to Paris. Flight-Lieutenant S. N. Webster, A.F.C., won the International Schneider Trophy for Great Britain at a speed of 281.656 m.p.h.

1928 H. J. Hinkler flew from England to Australia, covering 12,250 miles in $15\frac{1}{2}$ days.

'Southern Cross', a Fokker monoplane, with Captain C. Kingsford-Smith and C. Ulm, Australia, and H. W. Lyon and J. Warner, United States, flew from Oakland, California, to Sydney, Australia,

Southern Cross ▶

7,400 miles, stopping only at the Hawaiian Islands, Fiji Islands and Brisbane, Australia.

Juan de la Cierva of Spain flew by auto-giro from London to Paris.

1929   Airship *Graf Zeppelin* made a world tour.

Flying-Officer H. R. D. Waghorn, Great Britain, won the Schneider Seaplane Trophy race at an average speed of 328.64 m.p.h. Squadron-Leader H. Orlebar, Great Britain, set a world speed record of 357.7 m.p.h. at Calshot, England.

1930   Amy Johnson, Great Britain, completed 9,900 mile flight from London to Australia in $19\frac{1}{2}$ days.

Squadron-Leader Kingsford-Smith, with three companions, in the 'Southern Cross' monoplane, made the second east-west crossing of the Atlantic from Ireland to Newfoundland; 1,900 miles in 32 hours 12 minutes.

British airship R.100 flew from Carding-ton, England, to Montreal, Canada, in 78 hours 49 minutes. Returned in 57 hours.

1931   Professor Auguste Piccard and Dr. Charles Kipfer ascended to record height of 51,775 feet at Augsburg, Germany, in a balloon. Wiley Post, U.S. pilot, and Harold Gatty, Australian, as navigator, encircled the globe in 8 days 15 hours 51 minutes, a record-breaking 15,474-mile flight.

Flight-Lieutenant J. N. Boothman won the Schneider Trophy outright for Great Britain at a speed of 340.08 m.p.h.

1931   Flight-Lieutenant G. H. Stainforth, Great Britain, made new seaplane speed record of 406.997 m.p.h. at Calshot, England. H.J.L. ('Bert') Hinkler made first west-east crossing of South Atlantic in a 22-hour flight from Natal, Brazil, to Bathurst, British Gambia, Africa.

J. A. Mollison flew from Australia to England in 8 days 22 hours 25 minutes.

Schneider Trophy Seaplane ▶

1932 Miss Amelia Earhart flew from New-foundland to Ireland in $13\frac{1}{2}$ hours, the first solo flight across the Atlantic by a woman.

Auguste Piccard and Max Cosyns set altitude record of 53,153 feet in a balloon over Switzerland and Italy.

1933 Wiley Post, flying solo, encircled the globe in 7 days 18 hours 50 minutes, covering 15,596 miles.

1934 Major W. E. Kepner, Captain A. W. Stevens and Captain O. A. Anderson, taking off from Rapid City, South Dakota, ascended to 60,613 feet in a balloon. Lieutenant Francesco Agello made sea-plane record of 440.68 m.p.h. at Desenzano, Italy.

1935 Fred and Al Keys set a new refuelling endurance record over Meridian, Mississippi, by staying in the air 3 weeks 3 days 5 hours and 34 minutes.

1936 Amy Mollison flew from Lympne, England, to Cape Town, South Africa, in 3 days 6 hours 26 minutes, setting a new record.

Squadron-Leader F. R. D. Swain, Great Britain, set up new world's height record for landplanes of 49,967 feet.

1937 Three Russian airmen, M. Gromov, A. Yumachev and S. Danilin, made new long-distance record by flying from Moscow, Soviet Union, to San Jancinto, California, a distance of 6,296 miles.

1938 Two R.A.F. single-engined Vickers Wellesley long-range bombers established new world's distance record for landplanes of 7,162 miles by flying from Ismailia, Egypt, to Darwin, Australia.

1939 First commercial trans-Atlantic service begun by Pan American 'Yankee Clipper', flying by way of New York, Bermuda, Azores, Lisbon, Bordeaux, Marseilles and Southampton.

First British trans-Atlantic air mail service.

First jet plane, the German Heinkel He 178.

1941 First flight by British jet plane, the Gloster E.28/39.

1945 Group-Captain H. J. Wilson, Great Britain, in a Gloster Meteor, raised the world's speed record to 606 m.p.h. at Herne Bay, Kent.

1946 Group-Captain E. M. Donaldson, D.S.O., A.F.C., in a Gloster Meteor, raised the world's speed record to 616 m.p.h.

1947 Major Marion E. Carl, U.S. Marine Corps, flying a Douglas Skystreak jet plane, raised the world's speed record to 650.57 m.p.h.

1947 Captain Charles Yeager, United States Army Air Force, made the first supersonic flight in the rocket-powered Bell X-1 research aircraft.

1948 John Cunningham, Great Britain, set new altitude record for heavier-than-air craft in a de Haviland Vampire jet fighter of 59,445.5 feet at Hatfield, Hertfordshire.

Major Richard L. Johnson, United States Army Air Force, set a new world speed record by flying at 670.37 m.p.h.

1949 Squadron-Leader Trevor S. Wade, Great Britain, in a Hawker P.1052, flew from London to Paris in 20 minutes 37 seconds, an average speed of 618.26 m.p.h.

First flight of the 'Comet' jet airliner.

First flight of the Brabazon airliner.

1950 John Cunningham and crew, Great Britain, flew from London to Rome in the 'Comet' jet airliner in 1 hour 59 minutes 37 seconds, an average speed of 447 m.p.h.

1951 Roland P. Beamont, D.S.O., D.F.C., and crew, Great Britain, in an English Electric Canberra jet bomber, established a course record between Aldergrove, Belfast, and Gander Lake, Newfoundland, by flying the 2,072.04 statute miles in 4 hours 18 minutes 24.4 seconds at an average speed of 581.12 m.p.h.

LANDMARKS

1952 B.O.A.C. 'Comet' inaugurated world's flrst jet-liner service, London to Johannesburg. Flying time 16 hours 32 minutes, May 2. Returning in 16 hours 56 minutes.

1955 Entry into service of the first big jet—the Boeing 707.

1955 Colonel H. A. Hanes, U.S.A., set up unofficial speed record of 800 m.p.h. at Palmdale, California.

1956 Peter Twiss flew Fairey Delta 2 at a speed of 1,121 m.p.h. near Chichester.

1957 U.S.A.F. Stratojets make non-stop flight round world in 45 hours 19 minutes. Captain J. W. Kittinger made record balloon altitude flight of 96,000 feet. English Electric P.1 exceeded world speed record of 1,132 m.p.h.

1958 De Haviland Comet IV flew from New York to Hatfield in 6 hours 18 minutes at an average speed of 562 m.p.h.

1961 First flight of the revolutionary "Kestrel", world's first fixed-wing VTOL combat aircraft. On November 22nd, Lt.-Colonel Robert R. Robinson of the U.S. Marine Corps achieved a maximum speed of 1,606 m.p.h.

1963 First flight of the Boeing 727. Colonel R. L. Stephens, U.S.A.F., flew a Lockheed YF-12A at a speed of 2,070 m.p.h.

1968 The first flight of the world's first supersonic airliner, the Russian-built Tu 144, took place on December 31st.

1969 Concorde 001, the French prototype supersonic airliner flew for the first time on March 2nd at Toulouse and the British prototype Concorde 002, flew for the first time on April 9th at Filton. Regular flights of the Boeing 747—the jumbo-jet—between New York and London began from December.

BUILT BY
Wilbur and Orville Wright
DIMENSIONS
Length   21 ft. (6.4 m.)
Wing span   40 ft. (12.2 m.)
Weight empty
   605 lb. (275 kg.)
PASSENGERS/CARGO
1 pilot
SPEED
30 m.p.h. (48 km.p.h.)
CEILING
10 ft. (3m.) fully loaded
RANGE
852 ft. (260 m.)
ENGINE
One 12 h.p. Wright

The Flyer was the first powered, heavier-than-air aircraft to make a successful sustained and controlled flight. This historic flight was made on 17th December, 1903, at Kitty Hawk, with Orville Wright at the controls.

FLYER 1
1903

This was the first fully practical powered aeroplane. It made 59 flights in 1905, including one lasting over 38 minutes. In 1908 it was modified to carry a passenger, sitting on the leading edge of the wing.

FLYER 3
1905

BUILT BY
Wilbur and Orville Wright
DIMENSIONS
Length   28 ft. (8.5 m.)
Wing span   40 ft. (12 m.)
Weight empty
  710 lb. (322 kg.)
PASSENGERS/CARGO
1 pilot and 1 passenger
SPEED
35 m.p.h. (56 km.p.h.)
RANGE
24 miles (38 km.)
ENGINE
One 20 h.p. Wright

14-bis
1906

This grotesque machine is famous and remembered only because it made the first official aeroplane flights in Europe in 1906. This aircraft was a freak and did not help the development of practical aeroplanes. The peculiar name "14-bis" was given to it because Alberto Santos-Dumont first tested the machine slung under his dirigible airship No. 14.

BUILT BY
Santos-Dumont
DIMENSIONS
Wing span   38 ft. (11.5 m.)
Weight empty
    660 lb. (300 kg.)
PASSENGERS/CARGO
1 pilot
SPEED
25 m.p.h. (40 km.p.h.)
ENDURANCE
21 sec.
ENGINE
One 50 h.p. Antoinette

BUILT BY
Voisin
DIMENSIONS
Wing span   33 ft. (10.2 m.)
Weight empty
   115 lb. (520 kg.)
PASSENGERS/CARGO
1 pilot
SPEED
40 m.p.h. (64 km.p.h.)
RANGE
1,640 yds. (1,500 m.)
ENGINE
One 50 h.p. Antoinette

VOISIN-FARMAN 1
1907
18

Made by Voisin, this machine was modified by Henry Farman. In this form it became the first aeroplane, after the Wright Flyer No. 2, 1904, to remain airborne for over a minute and turn a complete circle. The latter flight, made on 13th January, 1908, won Farman the Deutsch-Archdeacon prize of 50,000 francs for completing the first official circle of one kilometre in diameter.

BUILT BY
A. V. Roe
DIMENSIONS
Length   23 ft. (7 m.)
Wing span   30 ft. (9 m.)
Weight empty
   600 lb. (272 kg.)
PASSENGERS/CARGO
1 pilot
RANGE
150 ft. (46 m.)
ENGINE
One 24 h.p. Antoinette

Tested initially by being towed behind a car at Brooklands, this purposeful biplane made short hops in 1908, including one believed to be 150 ft. (46 m.) in June. The flight was not officially observed, however, and did not allow Alliot Verdon Roe to claim he was the first Briton to fly a powered aeroplane in Britain.

ROE 1
1908

Considered the best of the early monoplanes, the Antoinette was a beautifully made machine and of modern configuration. An Antoinette just failed to become the first aeroplane to cross the English Channel when, during the attempt, its engine stopped a few miles from the coast. Latham, the pilot, ditched safely and was rescued by a destroyer.

ANTOINETTE 1909

BUILT BY
M. Levavasseur
DIMENSIONS
Length   37 ft. (11.5 m.)
Wing span   42 ft. (12.8 m.)
Weight empty
    992 lb. (450 kg.)
PASSENGERS/CARGO
1 pilot
SPEED
43 m.p.h. (70 km.p.h.)
RANGE
96 miles (155 km.)
ENGINE
One 50 h.p. Antoinette

BLÉRIOT XI
1909

BUILT BY
M. Blériot
DIMENSIONS
Length   26 ft. (8 m.)
Wing span   25 ft. (7.8 m.)
Weight empty
    460 lb. (210 kg.)
PASSENGERS/CARGO
1 pilot
SPEED
36 m.p.h. (58 km.p.h.)
RANGE
25+ miles (40+ km.) at
    100% maximum power
ENGINE
One 25 h.p. Anzani

Flown by M. Louis Blériot, the model XI was the first aircraft to cross the English Channel, doing so on 25th July, 1909. The venture nearly ended in disaster as, in mid-channel, the engine began to overheat. Fortunately a providential rain shower cooled it just in time. An improved version of this aircraft was used for reconnaissance duties in the First World War.

BUILT BY
G. Curtiss
DIMENSIONS
Length   28 ft. (8.5 m.)
Wing span   29 ft. (8.8 m.)
Weight empty
   550 lb. (250 kg.)
PASSENGERS/CARGO
1 pilot
SPEED
45 m.p.h. (72 km.p.h.)
ENGINE
One 50 h.p. Curtiss

GOLDEN FLYER
1909

Developed from the earlier *June Bug*, the *Golden Flyer* represented an important compromise between stability and controllability. It started the main rival biplane tradition to the Wright Flyer in the United States.

BUILT BY
M. Voisin
DIMENSIONS
Length   39 ft. (12 m.)
Wing span   33 ft. (10 m.)
Weight empty
      1,100 lb. (500 kg.)
PASSENGERS/CARGO
1 pilot
SPEED
34 m.p.h. (55 km.p.h.)
ENGINE
One 50 h.p. Antoinette

Voisin of the type used for the first officially recognised flight by a British pilot in Britain, which was made in April 1909 by J. T. C. Moore-Brabazon (later Lord Brabazon of Tara), flying his own Voisin named *Bird of Passage*.

**VOISIN
1909**

Relatively easy to fly, this aircraft became popular as a safe, sporting machine. It won the prize for the longest distance flown during the great air show at Rheims in 1909, which was attended by a quarter of a million people.

HENRY FARMAN III 1909

BUILT BY
Henry Farman
DIMENSIONS
Length   39 ft. (12 m.)
Wing span   33 ft. (10 m.)
Weight empty
    990 lb. (450 kg.)
PASSENGERS/CARGO
1 pilot
SPEED
37 m.p.h. (60 km.p.h.)
CEILING
360 ft. (110 m.) fully loaded
RANGE
112+ miles (180+ km.)
ENGINE
One 50 h.p. Gnome

MAURICE FARMAN "LONGHORN" 1911

BUILT BY
Henry and Maurice Farman,
  Billancourt
DIMENSIONS
Length   32 ft. (9.7 m.)
Wing span 51 ft. (15.5 m.)
Weight empty
  1,280 lb. (580 kg.)
PASSENGERS/CARGO
1 pilot and 1 observer
ARMAMENT
Rifles or duck guns carried
  by the observer
MAX. SPEED
59 m.p.h. (95 km.p.h.)
RANGE
350 miles (560 km.)
ENGINE
One 70 h.p. Renault, or one
  100 h.p. Sunbeam

In spite of its Wright-type forward elevator, which made it unstable and tricky to fly, the Longhorn was used extensively for training purposes, particularly in Britain. French squadrons used Longhorns for reconnaissance duties until 1915. A development of the Longhorn, with the elevator located behind the rudders, was appropriately known as the "Shorthorn".

BUILT BY
A. V. Roe, Manchester
DIMENSIONS
Length   29.5 ft. (9.0 m.)
Wing span   36 ft. (11.0 m.)
Weight empty
     1,100 lb. (500 kg.)
PASSENGERS/CARGO
1 pilot and 1 passenger
ARMAMENT
One fixed Lewis machine-gun
     (Home Defence aircraft)
MAX. SPEED
82 m.p.h. (132 km.p.h.)
CEILING
13,000 ft. (4,000 m.) fully
     loaded
ENDURANCE
3 hours
ENGINE
One 80 h.p. Gnome, 100 h.p.
     Gnome Monosoupape, 110
     h.p. Clerget, 110 h.p. Le
     Rhône, 75 h.p. Rolls-Royce
     Hawk, 130 h.p. Clerget

AVRO 504
1914

Over 8,000 of this two-seat trainer were built; it is generally considered to be the greatest training machine of all time. Many versions were produced, one of the best known being the 504J. Nearly 300 504s were fitted with machine-guns and served with Home Defence Units for use against Zeppelins and Gothas attempting to bomb London.

BUILT BY
Morane-Saulnier, Paris
DIMENSIONS
Length
  22.5 ft. (6.85 m.)
Wing span
  27.25 ft. (8.4 m.)
Weight empty
  735 lb. (333 kg.)
PASSENGERS/CARGO
1 pilot
ARMAMENT
One fixed machine gun, firing
  forward
MAX. SPEED
102 m.p.h. (164 km.p.h.)
CEILING
13,000 ft. (4,000 m.) fully
  loaded
RANGE
150 miles (240 km.)
ENGINE
One 80 h.p. Le Rhône, or one
  110 h.p. Le Rhône

This fighter, one of the best of its time, was in service when the First World War started. One was fitted with a crude "bullet deflector" gear enabling a forward firing machine-gun to be fitted, but was shot down and captured. Antony Fokker examined the gear, and realising its importance, designed and fitted a more efficient system to his Eindekker monoplanes, giving them superiority over Allied aircraft during 1915.

MORANE-SAULNIER SCOUT
1914   27

A two-seat trainer, the JN.4 was one of the most widely used training aeroplanes, being used throughout the 1914—18 war period in America, Canada, Britain and France. Many pupils crashed, however, when attempting aerobatics, for which the aircraft had not been designed.

JN.4 JENNY 1914

BUILT BY
Curtiss Aeroplane and Motors
  Corporation
DIMENSIONS
Length   27 ft. (8.2 m.)
Wing span   44 ft. (13.4 m.)
Weight empty
  1,580 lb. (716 kg.)
PASSENGERS/CARGO
1 pilot and 1 pupil
MAX. SPEED
70 m.p.h. (113 km.p.h.)
RANGE
200 miles (320 km.)
ENGINE
One 90 h.p. Curtiss OX-5, or
  one Hispano-Suiza

EINDEKKER SCOUT 1914

Known as the ''Fokker monoplane'', this aircraft ushered in the era of air combat, being fitted with Fokker's interrupter gear, which ''timed'' bullets so that they passed between the propeller blades, enabling pilots to aim by simply pointing the complete aircraft. Although only about 200 were built they were effective enough to gain supremacy over the Western Front in 1915 and early 1916.

BUILT BY
Fokker Flugzeugwerke,
   Schwerin
DIMENSIONS
Length
   24 ft. (7.3 m.)
Wing span
   31.25 ft. (9.5 m.)
Weight empty
   878 lb. (398 kg.)
PASSENGERS/CARGO
1 pilot
ARMAMENT
One or two fixed machine-
   guns, firing forward
MAX. SPEED
87 m.p.h. (140 km.p.h.)
CEILING
12,000 ft. (3,650 m.) fully
   loaded
ENDURANCE
1½ hours
ENGINE
One 100 h.p. Oberusel U-1

BUILT BY
Vickers, London
DIMENSIONS
Length
27.1 ft. (8.25 m.)
Wing span
36.5 ft. (11.1 m.)
Weight empty
1,220 lb. (553 kg.)
PASSENGERS/CARGO
1 pilot and 1 observer
ARMAMENT
One free machine-gun, plus
small bombs
MAX. SPEED
70 m.p.h. (113 km.p.h.)
CEILING
9,000 ft. (2,700 m.) fully
loaded
ENDURANCE
4½ hours
ENGINE
One 100 h.p. Gnome
Monosoupape

GUNBUS
1915
30

This two-seater fighter and reconnaissance aircraft helped to overcome the menace of the Fokker monoplanes. Although it was relatively slow, its forward mounted machine-gun proved very effective. When the aircraft went into service in 1914, it was the only one with provision for mounting a machine-gun—hence the name "Gunbus".

BUILT BY
Soc. Anonyme des Establis-
    sements Nieuport, Issy-
    le Molineux

DIMENSIONS
Length
    19.5 ft. (5.9 m.)
Wing span
    27.25 ft. (8.3 m.)
Weight empty
    825 lb. (374 kg.)

PASSENGERS/CARGO
1 pilot

ARMAMENT
One machine-gun, and eight
    Le Prieur rockets

MAX. SPEED
107 m.p.h. (172 km.p.h.)

CEILING
17,400 ft. (5,300 m.) fully
    loaded

ENDURANCE
2 hours

ENGINE
One 110 h.p. Le Rhône

This single-seater fighter helped end the ''Fokker Scourge'' of the Eindekker Scout. The improved model 17 entered service in 1916 and proved to be one of the best fighters of the First World War. When used for attacks on German observation balloons, four Le Prieur ''firework-type'' rockets were attached to each interplane strut, and were fired electrically.

**NIEUPORT 17
SCOUT
1915**

31

Known popularly as the ''225'' but officially as the Short Type 184, this naval aircraft was the only floatplane to take part in the battle of Jutland in 1916, when one was used for reconnaissance. The 225 was also the first aircraft to sink a ship by torpedo when it attacked a Turkish troopship during the Gallipoli campaign.

SHORT 225 SEAPLANE 1915

BUILT BY
Short Brothers, Rochester
DIMENSIONS
Length
   40.6 ft. (12.37 m.)
Wing span
   63.5 ft. (19.35 m.)
Weight empty
   3,703 lb. (1,679 kg.)
PASSENGERS/CARGO
1 pilot and 1 observer
ARMAMENT
One machine-gun, in rear
   cockpit
One 14-in: (35 c.m.) torpedo,
   or 520 lb. (240 kg.) bombs
MAX. SPEED
88 m.p.h. (142 km.p.h.)
CEILING
9,000 ft. (2,740 m.) fully
   loaded
ENDURANCE
2¾ hours
ENGINE
One 225 h.p. Sunbeam

L.V.G. BOMBER 1915

Used for reconnaissance and bombing, a small number of L.V.G.s were in the German air force at the outbreak of war in 1914. Like their British counterparts, they were unarmed and suffered heavy losses. This resulted in the much improved C.V, armed with machine-guns, which proved a formidable adversary, in spite of the pilot's exceptionally poor view forward.

BUILT BY
Luft Veherhs Gesellschaft, Johannistal

DIMENSIONS
Length
26.5 ft. (8.07 m.)
Wing span
42.75 ft. (13.02 m.)
Weight empty
1,860 lb. (843 kg.)

PASSENGERS/CARGO
1 pilot and 1 observer

ARMAMENT
One fixed machine-gun firing forward, one free gun

MAX. SPEED
102 m.p.h. (164 km.p.h.)

CEILING
16,500 ft. (5,000 m.) fully loaded

ENDURANCE
3½ hours

ENGINE
One 200 h.p. Benz BzIV

BUILT BY
Sopwith Aviation Co,
   Kingston-on-Thames
DIMENSIONS
Length
   19.25 ft. (5.9 m.)
Wing span
   26.5 ft. (8.1 m.)
Weight empty
   787 lb. (357 kg.)
PASSENGERS/CARGO
1 pilot
ARMAMENT
One fixed machine-gun
   firing forward
Four 25 lb. (11 kg.) bombs
MAX. SPEED
112 m.p.h. (180 km.p.h.)
CEILING
17,500 ft. (5,300 m.) fully
   loaded
ENDURANCE
3 hours
ENGINE
One 80 h.p. Le Rhône.

**PUP
1916
34**

A single-seat scout, the Pup is considered by many pilots to be the most pleasant aircraft to fly ever built. In combat it was so manoeuvrable that it could out-fly the formidable German Albatros fighter. Pups were used in pioneer carrier operations, an aircraft of this type making the first-ever landing on a ship under way.

This single-seat fighter was one of the more successful aircraft of the First World War; 8,472 were built. It was used extensively by French, Italian and American air units and, on the Western front, two British squadrons who flew machines "swopped" from the R.N.A.S. for Sopwith Triplanes. SPADs equipped the French "Cigognes" whose insignia was a symbolic stork, and who used the machine's ability to dive steeply without falling to bits to good effect in dog fights.

BUILT BY
Société Pour Aviation et des
    Denives, Paris
DIMENSIONS
Length   20.7 ft. (6.34 m.)
Wing span 27 ft. (8.23 m.)
Weight empty
    1,255 lb. (570 kg.)
PASSENGERS/CARGO
1 pilot
ARMAMENT
Two fixed machine-guns,
    firing forward
MAX. SPEED
130 m.p.h. (210 km.p.h.)
CEILING
22,000 ft. (6,700 m.) fully
    loaded
ENDURANCE
2 hours
ENGINE
One 200 h.p. Hispano-Suiza

SPAD SCOUT 1916

When this little seaplane took off for the first time nobody could have forseen that the 21 men who built it in an old boat-house would become the world's biggest aeroplane company. From the B & W developed both the Flying Fortress bombers of the Second World War and the Boeing jet airliners. The picture shows a flying scale replica of the original, built to commemorate Boeing's 50th anniversary. It has a good rate of climb—it can get up to 5000 ft. (1500 m.) in half an hour!

BOEING B & W SEAPLANE 1916

BUILT BY
Boeing Airplane Co, Seattle
DIMENSIONS
Length   27.5 ft. (8.38 m.)
Wing span 52 ft. (15.85 m.)
Weight
     2,800 lb. (1,270 kg.)
PASSENGERS/CARGO
1 pilot
MAX. SPEED
78 m.p.h. (125 km.p.h.)
RANGE
120 miles (190 km.)
ENGINE
One 125 h.p. Hall-Scott

Known affectionately as the "Brisfit", this was one of the war's outstanding combat aircraft, although it got off to a bad start when four were shot down by von Richthofen, because the pilots adopted the usual 2-seater defensive technique of letting their observer do most of the shooting. However, when pilots learned to fly the manoeuvrable Brisfit like a single-seater, front guns as the main weapon, observer's gun sting-in-the-tail, it won both victories and respect from the enemy.

BRISTOL FIGHTER 1916

BUILT BY
British and Colonial
    Aeroplane Co, Bristol
DIMENSIONS
Length   26 ft. (7.92 m.)
Wing span 39.25 ft. (12 m.)
Weight empty
    1,934 lb. (877 kg.)
PASSENGERS/CARGO
1 pilot and 1 observer
ARMAMENT
One fixed machine-gun,
    firing forward
One or two free machine-
    guns in the rear cockpit
Bomb load: 240 lb. (110 kg.)
MAX. SPEED
123 m.p.h. (200 km.p.h.)
CEILING
21,500 ft. (6,500 m.)
ENDURANCE
3 hours
ENGINE
One 275 h.p. Rolls-Royce
    Falcon III

BUILT BY
Aircraft Manufacturing Co,
  Hendon
DIMENSIONS
Length   30.7 ft. (9.23 m.)
Wing span 42.3 ft. (12.9 m.)
Weight empty
  2,303 lb. (1,044 kg.)
PASSENGERS/CARGO
1 Pilot and 1 observer
ARMAMENT
One fixed machine-gun firing
  forward (Standard version)
Two fixed machine-guns
  firing forward (R.N.A.S.
  aircraft)
One or two free guns in the
  rear cockpit
Max. bomb load
  460 lb. (210 kg.)
MAX. SPEED
119 m.p.h. (190 km.p.h.)
CEILING
16,000 ft. (4,900 m.)
ENGINE
One 250 h.p. Rolls-Royce
  Eagle III

D.H.4   1917
38

A day-bomber, the D.H.4 was fast enough to outfly enemy fighters, as did its successor, the
Mosquito in the Second World War. The wide gap between the two cockpits made com-
munication between the pilot and his observer difficult, but this handicap did not prevent
the D.H.4 from becoming the outstanding day bomber of the First World War. Altogether
1,449 were built.

BUILT BY
Albatros Werke, Berlin
DIMENSIONS
Length
    24 ft. (7.32 m.)
Wing span
    29.7 ft. (9.04 m.)
Weight empty
    1,457 lb. (660 kg.)
PASSENGERS/CARGO
1 pilot
ARMAMENTS
Two fixed machine-guns,
    firing forward
MAX. SPEED
103 m.p.h. (165 km.p.h.)
CEILING
18,000 ft. (5,500 km.)
    fully loaded
ENDURANCE
2 hours
ENGINE
One 170 h.p. Mercedes D.IIIa

To regain supremacy, after the ''Fokker Scourge'' had been overcome, the Germans put the Albatros series into production. The D.III, in units of 14 and led by aces, proved a formidable adversary, shooting down many British B.E.2c's. The Albatros had an unusually smooth and ''streamlined'' fuselage for its day and this undoubtedly contributed to its fine performance.

ALBATROS SCOUT
1917

The basic advantage claimed for this single-seater fighter was its wing arrangement; this obscured the pilot's view less than the broader chord wings of biplanes and monoplanes, and gave it a rapid rate of climb and a high degree of manoeuvrability. Although not so famous as its Fokker counterpart, the Sopwith Triplane achieved impressive success in its brief career, which lasted from early 1917 until the autumn of that year, when it was superseded by the Camel.

SOPWITH TRIPLANE 1917

BUILT BY
Sopwith Aviation Co, Kingston
DIMENSIONS
Length
18.8 ft. (5.72 m.)
Wing span
26.5 ft. (8.07 m.)
Weight empty
1,100 lb. (500 kg.)
PASSENGERS/CARGO
1 pilot
ARMAMENT
One fixed machine-gun, firing forward
MAX. SPEED
117 m.p.h. (190 km.p.h.)
CEILING
20,500 ft. (6,250 m.) fully loaded
ENDURANCE
2¾ hours
ENGINE
One 130 h.p. Clerget

This single-seater fighter, was, in combat efficiency, second only to the Sopwith Camel. Designed around the new French Hispano-Suiza engine it was slightly less manoeuvrable than the Camel, but steadier, making it a better gun platform for aerial fighting. Some famous British fighter aces scored many of their victories while flying S.E.5a's, including V.C. winners Mannock (73 victories), Bishop (72), McCudden (57).

S.E.5a
1917

BUILT BY
Royal Aircraft Factory,
  Farnborough
DIMENSIONS
Length
  21 ft. (6.4 m.)
Wing span
  26.6 ft. (8.1 m.)
Weight empty
  1,400 lb. (635 kg.)
PASSENGERS/CARGO
1 pilot
ARMAMENT
Two fixed machine-guns,
  firing forward, one on
  top of the fuselage, one
  above the wing centre
  section
MAX. CRUISING SPEED
138 m.p.h. (220 km.p.h.)
CEILING
17,000 ft. (5,200 m.) fully
  loaded
ENGINE
One 200 h.p. Hispano-Suiza

BUILT BY
Sopwith Aviation Co,
  Kingston
DIMENSIONS
Length
  18.75 ft. (5.7 m.)
Wing span
  28 ft. (8.53 m.)
Weight empty
  929 lb. (422 kg.)
PASSENGERS/CARGO
1 pilot
ARMAMENT
Two fixed machine-guns,
  firing forward
4 × 25 lb. (11 kg.) bombs
MAX. SPEED
115 m.p.h. (185 km.p.h.)
CEILING
19,000 ft. (5,800 m.) fully
  loaded
ENGINE
One 130 h.p. Clerget, or 150
  h.p. Bentley B.R.1

CAMEL
1917
42

The Camel which destroyed 1294 enemy aircraft, is considered the greatest fighter of the First World War. It was designed as a more powerful successor to the Pup and although not so pleasant to fly, it was even more manoeuvrable, and has been described as the most manoeuvrable aircraft ever built, being particularly fast on right hand turns, owing to the gyroscopic force produced by the rotary motor and the short fuselage. It was a tricky aircraft for novices, and many were killed attempting to make their first flight in one.

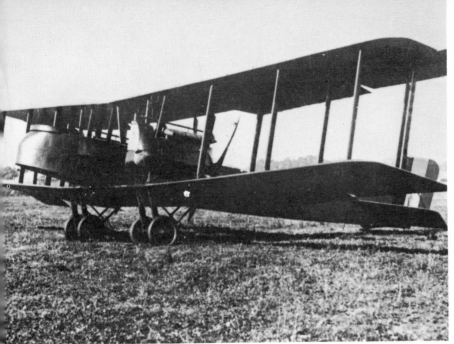

BUILT BY
Gothaer Waggonfabrik,
    Berlin

DIMENSIONS
Length
    39 ft. (11.89 m.)
Wing span
    77.75 ft. (23.67 m.)
Weight empty
    6,040 lb. (2,730 kg.)

PASSENGERS/CARGO
Crew: 3

ARMAMENT
Two machine-guns, one in
    nose and one in fuselage
Max. bomb load:
    1,100 lb. (500 kg.)

MAX. SPEED
87 m.p.h. (140 km.p.h.)

CEILING
20,500 ft. (6.250 m.)

RANGE
520 miles (840 km.)

ENGINES
2 × 260 h.p. Mercedes DIVa

This was the most successful of several long-range bombers designed by Germany. Gothas flew over London in daylight and dropped bombs with impunity on a surprised city only recently relieved of the threat of Zeppelin attacks. The Gotha threat, finally overcome by Bristols and Sopwith Camels, was on a scale out of all proportion to the small number of machines used and weight of bombs dropped.

**GOTHA
1917**

Britain's Royal Naval Air Service, the first air force to appreciate the damage that could be caused by bombing on a large scale, used the 0/400 to bomb Ostend, Zeebrugge and the aerodromes used by the Gothas which were attacking London. In 1918 0/400s were used for the first-ever sustained strategic night bombing offensive, dropping ''blockbusters'' weighing 1,650 lb. (750 kg.)

HANDLEY PAGE 0/400
1917

BUILT BY
Handley Page Ltd,
  Cricklewood
DIMENSIONS
Length
  62.8 ft. (19.1 m.)
Wing span
  100 ft. (30.48 m.)
Weight empty
  8,502 lb. (3,855 kg.)
PASSENGERS/CARGO
Crew: 3 or 4
ARMAMENT
Two or four machine-guns in
  nose and rear cockpits
Max. bomb load:
  2,000 lb. (900 kg.)
MAX. SPEED
97 m.p.h. (156 km.p.h.)
CEILING
8,500 ft. (2,600 m.) fully
  loaded
RANGE
770 miles (1240 km.)
ENGINES
2 × 250 h.p. Eagle IV

This reconnaissance flying boat is officially known as the H.12, but is better known as the "Large America" to distinguish it from its predecessor, the "Small America". Although the lightly constructed hull was easily damaged in rough seas, the Large America was extensively used by the R.N.A.S. for anti-submarine, anti-Zeppelin and general reconnaissance duties.

LARGE AMERICA 1917

BUILT BY
Curtiss Aeroplane and Motors
    Corporation, Hammondsport
DIMENSIONS
Length
    46.1 ft. (14.03 m.)
Wing span
    95 ft. (28.96 m.)
Weight empty
    7,360 lb. (3,340 kg.)
PASSENGERS/CARGO
Crew: 4
ARMAMENT
Five or six machine-guns
Bomb load:
    4 × 230 lb (100 kg.)
    bombs
MAX. SPEED
100 m.p.h. (160 km.p.h.)
CEILING
12,500 ft. (3,800 m.) fully
    loaded
ENGINES
2 × 345 h.p. Rolls-Royce
    Eagle

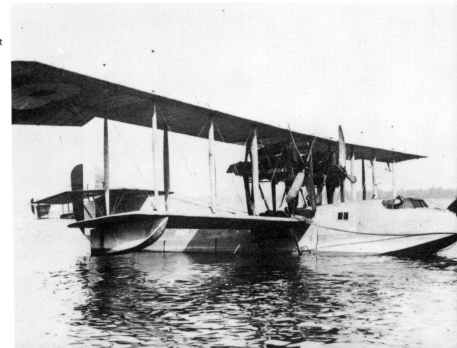

BUILT BY
Fokker Flugzeugwerke,
Schwerin

DIMENSIONS
Length
19 ft. (5.79 m.)
Wing span
23.6 ft. (7.19 m.)
Weight empty
893 lb. (405 kg.)

PASSENGERS/CARGO
1 pilot

ARMAMENT
Two fixed machine-guns,
firing forward

MAX. SPEED
115 m.p.h. (185 km.p.h.)

CEILING
19,600 ft. (6,000 m.) fully
loaded

ENDURANCE
1½ hours

ENGINE
One 110 h.p. Oberusel

FOKKER TRIPLANE 1917

46

This is the aircraft used in von Richthofen's circus. The Fokker's effectiveness was due to its rapid rate of climb, good manoeuvrability plus Allied confusion caused by its resemblance to the Sopwith Triplane, which is believed to have inspired it. Its weak structure, which caused several to break up in the air, quickly ended its spectacular career.

BUILT BY
Société des Moteurs
Salmson, Billancourt

DIMENSIONS
Length
28 ft. (8.53 m.)
Wing span
38.6 ft. (11.76 m.)
Weight empty
1,676 lb. (760 kg.)

PASSENGERS/CARGO
1 pilot and 1 observer

ARMAMENT
One fixed machine-gun,
firing forward
One or two free guns in the
rear cockpit
Bomb load: 500 lb (225 kg.)

MAX. CRUISING SPEED
115 m.p.h. (185 km.p.h.)

CEILING
20,500 ft. (6,250 m.) fully
loaded

ENGINE
One 260 h.p. Salmson

A two-seat reconnaissance aircraft, the Salmson was used as standard equipment in the American Expeditionary Force as well as with the French Air Force. In spite of its relatively large size, it could be a formidable adversary—one pilot shot down eight aircraft using his fixed, forward-firing machine-gun. The total number of Salmsons built was 3,200.

SALMSON
1917

47

This outstanding fighter helped to end the period of complete Allied superiority. Easy to fly, with a good rate of climb and a high ceiling, its main advantage was that it retained good control at heights where other aircraft were difficult to fly, and its performance was such that it was specifically named in the Armistice Agreement to be handed over to the Allies.

FOKKER D.VII 1918

BUILT BY
Fokker Flugzeugwerke,
    Schwerin
DIMENSIONS
Length
    23 ft. (7.01 m.)
Wing span
    29.25 ft. (8.91 m.)
Weight empty
    1,513 lb. (686 kg.)
PASSENGERS/CARGO
1 pilot
ARMAMENT
Two fixed machine-guns,
    firing forward
MAX. SPEED
124 m.p.h. (200 km.p.h.)
CEILING
22,900 ft. (7,000 m.) fully
    loaded
ENDURANCE
1½ hours
ENGINE
One 160 h.p. Mercedes D.III

A single-seat fighter, the Dolphin is noteworthy for the unusual back-stagger of its wings. In spite of a troublesome engine it proved itself effective for ground attack work, as well as aerial combat. A total of 1,532 Dolphins were built, and they remained in front-line service until the end of the war.

DOLPHIN
1918

BUILT BY
Sopwith Aviation Co,
   Kingston
DIMENSIONS
Length
   22.25 ft. (6.77 m.)
Wing span
   32.5 ft. (9.9 m.)
Weight empty
   1,406 lb. (638 kg.)
PASSENGERS/CARGO
1 pilot
ARMAMENT
Two fixed, forward firing
   machine-guns (standard)
Two additional guns could
   be mounted in the wing
Bomb load: 100 lb. (45 kg.)
MAX. CRUISING SPEED
128 m.p.h. (206 km.p.h.)
CEILING
21,000 ft. (6,400 m.) fully
   loaded
ENDURANCE
1¾ hours

BUILT BY
Sopwith Aviation Co,
   Kingston
DIMENSIONS
Length
   28.5 ft. (8.68 m.)
Wing span
   46.75 ft. (14.24 m.)
Weight empty
   2,199 lb. (997 kg.)
PASSENGERS/CARGO
1 pilot
ARMAMENT
One 18 in. (45 cm.) torpedo
MAX. SPEED
103 m.p.h. (166 km.p.h.)
CEILING
12,100 ft. (3,700 m.) fully
   loaded
ENDURANCE
4 hours
ENGINE
One 200 h.p. Sunbeam Arab

CUCKOO
1918

A single-seat torpedo bomber, the Cuckoo was so named because it was intended to lay its ''egg'' in other peoples' nests! The first squadron embarked on the carrier H.M.S. Argus in October, 1918, but the war ended before it could go into action. The aircraft is, however, credited with pioneering the techniques which were used so successfully during the Second World War.

BUILT BY
Vickers, London
DIMENSIONS
Length
    43.5 ft. (13.2 m.)
Wing span
    67.1 ft. (20.5 m.)
Weight empty
    7,100 lb. (3,220 kg.)
PASSENGERS/CARGO
Crew: 4
ARMAMENT
Two machine-guns, one each
    in nose and aft cockpits
Bomb load: 18 × 112 lb
    (50 kg.) bombs
2 × 230 lb (104 kg.) bombs
MAX. SPEED
103 m.p.h. (166 km.p.h.)
CEILING
7,000 ft. (2,100 m.)
RANGE
1,000 miles (1,600 km.)
ENGINES 2 × 360 h.p.
    Rolls-Royce VIII

VIMY
1918

This twin-engined bomber, too late for operational service in the First World War, achieved fame with a series of post-war long distance flights, most notably the first non-stop transatlantic flight by Alcock and Brown in June, 1919. Another, flown by the Smith Brothers and two mechanics, left Britain on 12th November, 1919 and reached Port Darwin Australia, 11,000 miles (17,700 km.) away, on 10th December. The Eagle was the most widely used of the various Vimy engines.

51

After the "sticks, wire and canvas" aircraft of the war, designers began metal construction, although Britain in particular tended merely to redesign biplanes of the wood era in metal. An exception was the graceful and ambitious Silver Streak, with a duralumin stressed-skin fuselage and wings, but this method was not generally adopted until the 1930s.

SILVER STREAK 1920

BUILT BY
Short Brothers, Rochester
DIMENSIONS
Length
    26.4 ft. (8.04 m.)
Wing span
    37.5 ft. (11.43 m.)
Weight empty
    1,865 lb. (846 kg.)
PASSENGERS/CARGO
1 pilot
CRUISING SPEED
90 m.p.h. (145 km.p.h.)
RANGE
450 miles (725 km.)
ENGINE
One 240 h.p. Siddeley Puma

RYAN M-1 MONOPLANE 1925

This combination mail and passenger carrier is chiefly remembered because of the special version, *Spirit of St. Louis* in which Lindbergh made his solo transatlantic flight in 1927. It was fitted with additional fuel tanks and navigational equipment, and Lindbergh covered the 3,600 miles (5,800 km.) from New York to Paris in 33 hours 39 minutes at an average speed of 107 m.p.h. (171 km.p.h.). The flight helped to make America air-minded and gain her her present lead in transport aircraft.

BUILT BY
B. F. Mahoney Aircraft
    Corporation, San Diego
DIMENSIONS
Wing span   46 ft. (14 m.)
Weight empty
    2,000 lb. (907 kg.)
PASSENGERS/CARGO
1 pilot; 2 passengers
CRUISING SPEED
105 m.p.h. (170 km.p.h.)
CEILING
17,500 ft. (5,300 m.) fully
    loaded
ENGINE
One 200 h.p. Wright
    Whirlwind

BUILT BY
Fokker, Schipol
DIMENSIONS
Length
    47.9 ft. (14.6 m.)
Wing span
    63.25 ft. (19.3 m.)
Weight empty
    4,730 lb. (2,145 kg.)
PASSENGERS/CARGO
2 crew; 8 passengers
MAX. CRUISING SPEED
106 m.p.h. (170 km.p.h.)
CEILING
15,500 ft. (4,700 m.) fully
    loaded
RANGE
500 miles (800 km.)
ENGINES
3 × 215 h.p. Armstrong
    Siddeley Lynx

FOKKER TRIMOTOR 1926

54

The best known of the Fokker monoplane transports which formed the backbone of European airlines for almost twenty years after the First World War is the Trimotor V1 1A/3M, which could remain airborne if any one of its three motors failed. Its enviable safety record resulted in a long period of service, during which Trimotors were used for long distance record attempts. One, *The Spider*, gave joy rides to nearly 40,000 people in six months.

BUILT BY
The Lockheed Aircraft
  Corporation, California
DIMENSIONS
Length
  27.5 ft. (8.37 m.)
Wing span
  41 ft. (12.5 m.)
Weight empty
  2,050 lb. (930 kg.)
PASSENGERS/CARGO
1 pilot; 6 passengers
CRUISING SPEED
135 m.p.h. (218 km.p.h.)
CEILING
17,250 ft. (5,250 m.) fully
  loaded
RANGE
900 miles (1,450 km.)
ENGINE
One 425 h.p. Pratt & Whitney
  Wasp

Several small but useful high winged, single-engined airliners carrying about six passengers at 100 m.p.h. appeared in the mid-1920s. The fashion for the type was set by the Lockheed Vega, with fully cantilevered wooden wings and finely streamlined wooden stressed-skin fuselage, made in two halves and glued together, plastic-model fashion. It was copied all over the world.

VEGA
1927

Typical of the biplane fighters of the time, the PW-9C was the last of a successful series of PW machines produced by Boeing. The PW-9C formed the U.S. Army Air Corps standard single-seaters for many years. The letters PW stand for "Pursuit-Watercooled", referring to the method of cooling employed on the engine.

BOEING PW-9C
1927

BUILT BY
Boeing Airplane Co, Seattle
DIMENSIONS
Length
   23.6 ft. (7.2 m.)
Wing span
   32 ft. (9.75 m.)
Weight
   3,043 lb. (1,380 kg.)
PASSENGERS/CARGO
1 pilot
MAX. SPEED
163 m.p.h. (262 km.p.h.)
CEILING
20,000 ft. (6,000 m.) fully
   loaded
ENGINE
One 400 h.p. Curtiss D.12

GIPSY MOTH 1928

This delightful little biplane extended the pleasure of flying almost universally. By 1929, eighty-five per cent of the privately-owned aircraft on the British register were Moths. Many famous long-distance flights were made in Moths, and height and speed records were also established by the pilots of these aircraft.

BUILT BY
The de Havilland Aircraft Co, Hatfield

DIMENSIONS
Length
  23.9 ft. (7.28 m.)
Wing span
  30 ft. (9.14 m.)
Weight
  1,750 lb. (793 kg.)

PASSENGERS/CARGO
1 pilot and 1 passenger

MAX SPEED
100 m.p.h. (160 km.p.h.)

ENGINE
One 100 h.p. D.H. Gipsy One

BUILT BY
Dornier-Flugzeuge,
   Altenrhein
DIMENSIONS
Length
   131.3 ft. (40.05 m.)
Wing span
   157.4 ft. (48 m.)
Weight empty
   64,900 lb. (29, 500 kg.)
PASSENGERS/CARGO
10 crew; 70–80 passengers
CRUISING SPEED
118 m.p.h. (190 km.p.h.)
CEILING
1,640 ft. (500 m.) fully
   loaded
ENGINE
12 × 600 h.p. Curtiss
Conqueror

DORNIER Do.X
1929
58

The hull of this twelve-engined flyingboat contained three decks: top, with the crew, engineer and radio compartments; centre, the 80 ft. (24.4 m.) long by 10 ft. (3 m.) wide passenger cabin, divided into lounges; bottom, fuel tanks and stores. Although for one demonstration a Do.X carried 170 passengers and crew, it was too big for its time, and woefully underpowered. Successive flying-boats were much smaller.

BUILT BY
The de Havilland Aircraft Co,
   Hatfield
DIMENSIONS
Length
   23.9 ft. (7.28 m.)
Wing span
   29.3 ft. (8.93 m.)
Weight
   1,825 lb. (827 kg.)
PASSENGERS/CARGO
1 pilot and 1 passenger
MAX. CRUISING SPEED
90 m.p.h. (145 km.p.h.)
ENGINE
One 130 h.p. D.H. Gipsy
   Major

One of the most famous and best loved aircraft of all time, the Tiger Moth has been called the ''trainer of the Empire''. Embodying many improvements on its model, the Gipsy Moth, the Tiger was superb in training (both elementary and aerobatic) and instruction in blind-flying, navigation, bombing, photography and even fighting tactics. 8,300 were built, it was adopted by over 25 countries, and was the R.A.F.'s primary trainer in the Second World War.

TIGER MOTH
1931

Although dated and slow even for its time, this stately and reliable aircraft, the most famous British pre-war airliner, carried up to 40 passengers in Pullman-class comfort. The eight built carried more passengers between London and the Continent than all other airliners combined, flew more than ten million miles (16 million km.) without hurting a passenger and were in the air for over 100,000 hours. Five also gave good service to the R.A.F. during the Second World War.

H.P. 42
1931

BUILT BY
Handley Page, Cricklewood
DIMENSIONS
Length
89.75 ft. (27.36 m.)
Wing span
130 ft. (39.6 m.)
Weight empty
21,000 lb. (9,620 kg.)
PASSENGERS/CARGO
2 crew; 40 passengers
MAX. CRUISING SPEED
100 m.p.h. (160 km.p.h.)
ENGINES
4 × 550 h.p. Bristol Jupiter

This single-seat interceptor-fighter was the first in squadron service to exceed 200 m.p.h. (320 km.p.h.), had a high rate of climb, exceptional manoeuvrability, and yet was easy to fly. This made it good for "air shows" and Furies thrilled spectators with their tied-together aerobatics. The Fury II of 1936, with the more powerful 640 h.p. Rolls-Royce Kestrel VI, was even more striking, and could be recognised by the spats fitted to its wheels.

FURY I
1931

BUILT BY
Hawker Aircraft, Kingston
DIMENSIONS
Length
 26.7 ft. (8.1 m.)
Wing span
 30 ft. (9.1 m.)
Weight empty
 2,623 lb. (1,190 kg.)
PASSENGERS/CARGO
1 pilot
ARMAMENT
Two synchronised machine-
 guns
MAX. SPEED
207 m.p.h. (333 km.p.h.)
CEILING
28,000 ft. (8,500 m.) fully
 loaded
RANGE
305 miles (490 km.)
ENGINE
One 525 h.p. Rolls-Royce
 Kestrel IIS

Junkers Flugzeug und
    Motorenwerke, Dassau
DIMENSIONS
Length
    62 ft. (18.9 m.)
Wing span
    95.9 ft. (29.2 m.)
Weight empty
    14,325 lb. (6,500 kg.)
PASSENGERS/CARGO
2 pilots; 17 troops
ARMAMENT
up to 4 machine-guns (Troop
    transport)
MAX. CRUISING SPEED
160 m.p.h. (260 km.p.h.)
CEILING
19,000 ft. (5,800 m.) fully
    loaded
RANGE
550 miles (885 km.)
ENGINES
3 × 770 h.p. BMW 132H

JUNKERS Ju.52
1932

62

This was a three-engined aeroplane, for troop, passenger or freight transport. Although civil versions were moderately successful, the type is best remembered as the aerial work-horse of the German Army during the Second World War. Large flaps gave low take-off and landing speeds, and a simple structure coupled with a distinctive corrugated metal skin imparted a robustness which became legendary.

DESIGNED BY
M. Mignet
DIMENSIONS
Length   13 ft. (3.96 m.)
Wing span   22 ft. (6.7 m.)
Weight empty
     350 lb. (160 kg.)
PASSENGERS/CARGO
1 pilot
MAX. CRUISING SPEED
60 m.p.h. (96 km.p.h.)
RANGE
200 miles (320 km.)
ENGINE
One, various makes, 22 to
     38 h.p.

FLYING FLEA
1933

This was intended to be a small, cheap do-it-yourself aircraft that almost any handyman could build from Mignet's design and then teach himself to fly. For simplicity the top wing pivoted about the front spar and was tilted up and down for longitudinal control. There were no ailerons, turns being made by the rudder alone. About 100 were built in England but accidents showed that the aircraft became uncontrollable if the wing incidence exceeded 15°, and the Fleas were banned.

This was the first modern type of airliner and introduced basic features still found on aircraft today. These are: monoplane configuration, two powerful engines, allowing level flight to be maintained if one should fail, low-wing all-metal, stressed-skin construction, variable-pitch propellers and retractable landing gear. However, its rival, the Douglas D.C.1. embodied wing flaps which rendered it superior.

BOEING 247
1933

BUILT BY
Boeing Airplane Co, Seattle
DIMENSIONS
Length
    51.3 ft. (16.25 m.)
Wing span
    74 ft. (22.6 m.)
Weight empty
    8,340 lb. (3,780 kg.)
PASSENGERS/CARGO
2 pilots, stewardess and 10
    passengers
MAX. CRUISING SPEED
171 m.p.h. (275 km.p.h.)
CEILING
18,500 ft. (5,650 km.) fully
    loaded
ENGINES
2 × 550 h.p. Pratt &
    Whitney Wasp

COMET RACER
1934

BUILT BY
The de Havilland Aircraft Co,
  Hatfield
DIMENSIONS
Length   29 ft. (8.84 m.)
Wing span   44 ft. (13.4 m.)
Weight max
   5,550 lb. (2,517 kg.)
PASSENGERS/CARGO
1 pilot and 1 navigator
MAX. SPEED
237 m.p.h. (380 km.p.h.)
RANGE
3,000 miles (4,800 km.)
ENGINES
2 × 230 h.p. D.H. Gipsy
  Six R

Designed for the great London-to-Melbourne Race of 1934, this was the first British aircraft to combine the three "modern features" of a variable pitch propeller, flaps and a retractable landing gear. Sixty-four aircraft were entered for the race; the Comet won (in 70hr. 54 minutes) because of its aerodynamic-cleanliness and efficiency, and demonstrated, like the DC-2, which came in second, the economic merits of the monoplane configuration for transport.

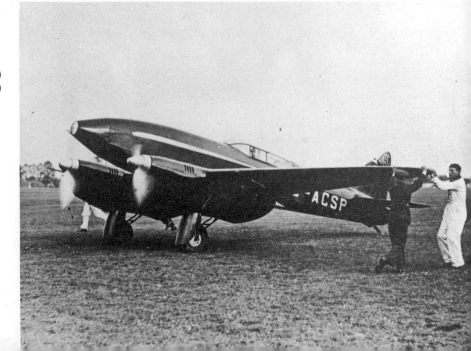

BUILT BY
Lockheed Aircraft
    Corporation, Burbank
DIMENSIONS
Length
    38.6 ft. (11.8 m.)
Wing span
    55 ft. (16.76 m.)
Weight empty
    6,325 lb. (2,868 kg.)
PASSENGERS/CARGO
2 pilots; 10 passengers
MAX. CRUISING SPEED
185 m.p.h. (300 km.p.h.)
CEILING
21,000 ft. (6,400 m.) fully
    loaded
RANGE
810 miles (1,300 km.)
ENGINES
2 × 450 h.p. Pratt &
    Whitney Wasp Junior

**ELECTRA
1934
66**

Following the Boeing 247 and starting another line of successful transports, the Electra gained a high reputation for speed and economy with American airlines, and in 1936 several were purchased by British Airways for use on its European routes, particularly to Scandinavia. It revolutionised travel between Britain and Europe's capital cities, helping to lay the foundations of today's services.

BUILT BY
The de Havilland Aircraft Co,
   Hatfield
DIMENSIONS
Length
   34.5 ft. (10.5 m.)
Wing span
   48 ft. (14.6 m.)
Weight empty
   3,230 lb. (1,464 kg.)
PASSENGERS/CARGO
2 crew; 8 passengers
CRUISING SPEED
132 m.p.h. (212 km.p.h.)
CEILING
19,500 ft. (5,940 m.) fully
   loaded
RANGE
580 miles (930 km.)
ENGINES
2 × 200 h.p. D.H. Gipsy
   Six, or Gipsy Queen

The D.H.89 Dragon Rapide, a simple, rugged and cheap twin-engined airliner, helped to introduce air travel to several remote areas. Some were used as personal run-abouts or executive aircraft for official duties.

Rapides serving with the R.A.F. were known as "Dominies", (a name often applied to civil Rapides also) and proved ideal for navigation and wireless training.

RAPIDE
1934

68

Affectionately known as "Stringbag", this was a more effective torpedo-spotter reconnaissance aircraft than its appearance suggests, largely due to its rugged construction and superb low-speed flying qualities, enabling it to take-off from and land on carriers pitching in rough seas. One of its exploits was to slow down the German *Bismarck* so that this formidable warship could be engaged and sunk by the Royal Navy.

SWORDFISH 1934

BUILT BY
Fairey Aviation Co, Hayes
DIMENSIONS
Length
    36.3 ft. (11 m.)
Wing span
    45.5 ft. (13.8 m.)
Weight empty
    5,200 lb. (2,360 kg.)
PASSENGERS/CARGO
Crew: 2 or 3
ARMAMENT
Two machine-guns
One 1,610 lb. (730 kg.)
    torpedo, or 8 × 60 lb (27
    kg.) rockets
MAX. SPEED
139 m.p.h. (224 km.p.h.)
CEILING
10,700 ft. (3,260 m.) fully
    loaded
RANGE
550 miles (885 km.)
ENGINE
One 690 Bristol Pegasus

DORNIER Do.215
1934

Nicknamed the "Flying Pencil" because of its slim fuselage, this was one of the main German bombers of the Second World War. Actually the export version of the Do.17 bomber which was based on the Do.17VI, a fast mailplane for the Deutsch Luft Hansa airline, it proved faster than most fighters of the day and was used by the Luftwaffe both as bomber and night fighter.

BUILT BY
Dornier-Flugzeuge,
   Altenrhein
 DIMENSIONS
Length
   51.75 ft. (15.75 m.)
Wing span
   59.1 ft. (18 m.)
Weight empty
   12,730 lb. (5,770 kg.)
PASSENGERS/CARGO
Crew: 4
ARMAMENT
Four machine-guns
Max. bomb load: 2,200 lb.
   (1,000 kg.)
MAX. SPEED
300 m.p.h. (480 km.p.h.)
CEILING
29,000 ft. (8,800 m.) fully
   loaded
RANGE
1,860 miles (3,000 km.)
ENGINES
2 × 1,075 h.p. Daimler-Benz

BUILT BY
Percival Aircraft, Gravesend
DIMENSIONS
Length
  18.23 ft. (5.56 m.)
Wing span
  24 ft. (7.32 m.)
Weight empty
  1,000 lb. (450 kg.)
PASSENGERS/CARGO
1 pilot
MAX. SPEED
195 m.p.h. (314 km.p.h.)
RANGE
550 miles (885 km.)
ENGINE
One 165 h.p. Napier Javelin

MEW GULL
1934
70

This clean, low-winged monoplane, with a finely streamlined cockpit cover and a spatted undercarriage, was designed for racing and six were built. They took part in many notable races and long-distance contests, including the 1939 flight from England to the Cape and back, which took 4 days, 10 hours, 16 minutes in a Gull with long-range tanks, flown by Alex Henshaw. The surviving Gull won several post-war King's Cup Air Races.

BUILT BY
Boeing Airplane Co, Seattle
DIMENSIONS
Length
  73.9 ft. (22.5 m.)
Wing span
  103.75 ft. (31.6 m.)
Weight empty
  32,250 lb. (14,630 kg.)
PASSENGERS/CARGO
Crew: 10
ARMAMENT
Up to thirteen 0.30 and 0.50
  in. machine-guns
Max. bomb load: 9,600 lb.
  (4,350 kg.)
MAX. SPEED
317 m.p.h. (510 km.p.h.)
RANGE
2,000 miles (3,220 km.) with
  4,000 lb. of bombs
ENGINES
4 × 1,200 h. p. Wright
R-1820

Mass production of this four-engined, heavy day-bomber did not start until 1939. Over
12,000 had been built by 1945. When first used against heavily-defended targets in
Europe, losses were high; therefore more defensive guns were added until the aircraft
richly deserved its popular name. They flew high in huge formations, protecting each
other with cross-fire, and dropped 640,036 tons of bombs in Europe for a loss of 4,750
aircraft.

FLYING FORTRESS
1935

This single-seat fighter is one of the greatest ever warplanes. More than 33,000 were built before and during the Second World War, more than any aircraft in history. Production continued in post-war Spain and Czechoslovakia and they were still in service 30 years after the prototype first flew, an all-time record. The main German fighter in the Battle of Britain, it was only slightly inferior to the British Spitfire.

MESSERSCHMITT Bf. 109E 1935

BUILT BY
Messerschmitt A.G.,
　Augsberg
DIMENSIONS
Length
　28.5 ft. (8.68 m.)
Wing span
　32.3 ft. (9.84 m.)
Weight empty
　4,420 lb. (2,005 kg.)
PASSENGERS/CARGO
1 pilot
ARMAMENT
3 × 20 mm. cannon and
　2 × 7.9 mm machine-
　guns
MAX. SPEED
354 m.p.h. (570 km.p.h.)
CEILING
36,000 ft. (11,000 m.)
RANGE
412 miles (660 km.)
ENGINE
One 1,100 h.p. Daimler-Ben
　DB 601A

DOUGLAS DC-3
1935

BUILT BY
The Douglas Aircraft
  Company, Santa Monica
DIMENSIONS
Length
  64.5 ft. (19.6 m.)
Wing span
  95 ft. (29 m.)
Weight empty
  16,290 lb. (7,390 kg.)
PASSENGERS/CARGO
2 crew; 21 passengers
MAX. CRUISING SPEED
185 m.p.h. (298 km.p.h.)
CEILING
23,000 ft. (7,000 km.)
RANGE
1,500 miles (2,400 km.)
ENGINES
2 × 900 h.p. Wright
  Cyclones

The most famous and successful airliner ever built, the DC-3, or ''Dakota'' revolutionised air transport, achieving success so quickly that by 1938 it was carrying the majority of American air travellers. Since then it has been used universally. In wartime it has carried troops and in Vietnam, fitted with rapid-fire, rotating-barrel Vulcan cannons, had the greatest firepower of any combat aircraft. Over 13,000 had been built when production ceased in 1945.

BUILT BY
Junkers Flugzeug and
   Motorenwerke, Dassau
DIMENSIONS
Length
   35.6 ft. (10.85 m.)
Wing span
   45.3 ft. (13.8 m.)
Weight empty
   6,085 lb. (2,760 kg.)
PASSENGERS/CARGO
1 pilot and 1 gunner
ARMAMENT
Two machine-guns in wings,
   one in rear cockpit
Bomb load: 1,100 lb
   (500 kg.)
MAX. SPEED
232 m.p.h. (373 km.p.h.)
RANGE
370 miles (600 km.)
ENGINE
One 1,100 h.p. Junkers
   Jumo 211

JUNKERS Ju.87B
1935

This dive-bomber was used effectively to help clear the path for the German invasion of Poland and other European countries in the Second World War. "Whistles" were fitted to the aircraft to increase its noise and power to terrify during the steep dive to drop its bombs. The R.A.F. shot it out of the skies when it used this tactic: it was only good where opposition was poor.

This four-seat executive monoplane, officially the Bristol Type 142, specially ordered by Lord Rothermere, the newspaper magnate, was found to be nearly 100 m.p.h. (160 km.p.h.) faster than the biplane fighters then with the R.A.F. Lord Rothermere named it *Britain First*, presented it to the nation, and from it was developed the Blenheim light bomber, for several years the R.A.F.'s fastest.

BUILT BY
The Bristol Aeroplane Co.,
   Bristol
DIMENSIONS
Length
   93.8 ft. (28.6 m.)
Wing span
   56 ft. (17.07 m.)
Weight
   9,812 lb. (4,450 kg.)
PASSENGERS/CARGO
2 crew; 4 passengers
MAX. SPEED
307 m.p.h. (493 km.p.h.)
ENGINES
2 × 560 h.p. Bristol Mercury
   VIS

BRITAIN FIRST
1935

Successfully designed as a simple, rugged and easy-to-fly light aeroplane for private owners, the Cub was cheap and could be flown from small airfields. This with a low landing speed of only 30 m.p.h. (50 km.p.h.) spelt instant success and over 760 had been sold by the end of 1936. Different engines were fitted to the later, widely exported models, which helped America gain her present lead in light aircraft.

PIPER CUB 1935

BUILT BY
Taylor Aircraft, Pennsylvania
DIMENSIONS
Length
    22.5 ft. (6.86 m.)
Wing span
    35.1 ft. (10.7 m.)
Weight empty
    560 lb. (254 kg.)
MAX. SPEED
87 m.p.h. (140 km.p.h.)
CEILING
12,000 ft. (3,650 km.) fully
    loaded
RANGE
210 miles (340 km.)
ENGINE
One 40 h.p. Continental
    A.404

HEINKEL He.111
1935

First used in the Spanish Civil War, the He.111 was one of the main bombers used against Britain during the Battle of Britain in 1940. The prototype flew in 1935—camouflaged as an airliner and fitted with seats for ten passengers. Used throughout the Second World War, some versions were adapted to drop torpedoes and also air-launch V-1 flying bombs.

BUILT BY
Ernst Heinkel Flugzeugwerke,
  Rostock
DIMENSIONS
Length
  54.6 ft. (16.63 m.)
Wing span
  74.1 ft. (22.6 m.)
Weight empty
  14,400 lb. (6,530 kg.)
PASSENGERS/CARGO
Crew: 5
ARMAMENT
1 × 20 mm. cannon
6 × 7.9 mm machine-guns
Max bomb load: 4,400 lb.
  (2,000 kg.)
MAX. SPEED
267 m.p.h. (430 km.p.h.)
RANGE
2,175 miles (3,500 km.)
ENGINES
2 × 1,200 h.p. Junkers
  Jumo 211D

BUILT BY
The Heston Aircraft Co,
    Hounslow
DIMENSIONS
Length
    30.1 ft. (9.15 m.)
Wing span
    40.3 ft. (12.3 m.)
Weight empty
    2,000 lb. (900 kg.)
PASSENGERS/CARGO
1 pilot; 4 passengers
MAX. CRUISING SPEED
135 m.p.h. (217 km.p.h.)
CEILING
20,000 ft. (6,000 m.) fully
    loaded
RANGE
500 miles (800 km.)
ENGINE
One 200 h.p. D.H. Gipsy VI

PHOENIX
1935

This five-seat training aircraft is best remembered for its revolutionary hydraulically-operated retractable landing gear. The main wheels retracted neatly into a small stub-wing which formed part of the wing-bracing struts. The aircraft had a particularly quiet cabin which was also bigger than normal for its type. It was not, however, commercially successful.

BUILT BY
Short Brothers, Rochester
DIMENSIONS
Length
88 ft. (26.82 m.)
Wing span
114 ft. (34.74 m.)
Weight empty
23,500 lb. (10,660 kg.)
PASSENGERS/CARGO
5 crew; 24 passengers
MAX. CRUISING SPEED
165 m.p.h. (265 km.p.h.)
CEILING
20,000 ft. (6,000 m.) fully
loaded
RANGE
800 miles (1,300 km.)
ENGINES
4 × 910 h.p. Bristol Pegasus

When it appeared, this flying-boat set a new style for its kind. Though slower than land-planes of comparable size, it was popular with passengers because of its comfort and safety, giving Britain a leadership on the civil air routes of the world unknown before or since that time. In eleven years the fleet covered nearly 40 million miles (64 million km.), over 2 million miles (3 million km.) by the *Canopus* alone.

**EMPIRE
1936**

Until the 4-engined heavy bombers appeared, the twin-engined Wellington was the mainstay of the British offensive in the Second World War. A novel feature was its immensely strong geodetic construction, consisting of a basket-like structure covered with fabric, and Wellingtons survived damage which would have downed more conventional aircraft. Over 11,000 were built and these dropped 42,440 tons of bombs on sorties from Britain, including the first 4,000 lb. (900 kg.) block-busters.

WELLINGTON 1936

BUILT BY
Vickers-Armstrong,
  Weybridge
DIMENSIONS
Length
  64.6 ft. (19.7 m.)
Wing span
  86.1 ft. (26.2 m.)
Weight empty
  15,887 lb. (7,233 kg.)
PASSENGERS/CARGO
Crew: 6
ARMAMENT
Eight machine-guns
Max. bomb load: 4,500 lb.
  (2,000 kg.)
MAX. SPEED
255 m.p.h. (410 km.p.h.)
CEILING
22,000 ft. (6,700 m.) fully
  loaded
RANGE
1,470 miles (2,365 km.)
ENGINES
2 × 1,590 h.p. Bristol
  Hercules (Mk III)

This four-seat reconnaissance bomber was one of Germany's most versatile and successful Second World War combat aircraft. It was used both as a day and night bomber, and a long-range and night fighter, the latter armed with cannon and often equipped with radar. The Ju.388 was produced to combat the R.A.F.'s high-flying Mosquito, and like the Ju.188 had more powerful engines and improved performance.

JUNKERS Ju.88
1936

BUILT BY
Junkers Flugzeug und
   Motorenwerke, Dassau
DIMENSIONS
Length
   47.1 ft. (14.33 m.)
Wing span
   65.8 ft. (20.1 m.)
Weight empty
   19,510 lb. (8,850 kg.)
PASSENGERS/CARGO
Crew: 4
ARMAMENT
Up to five machine-guns
Max. bomb load:
   4,190 lb. (1,900 kg.)
MAX. SPEED
300 m.p.h. (480 km.p.h.)
RANGE
2,980 miles (4,795 km.)
ENGINES
2 × 1,410 h.p. Junkers
   Jumo 211J

BUILT BY
General Aircraft Ltd, Feltham

DIMENSIONS
Length
  23.25 ft. (7.08 m.)
Wing span
  34.5 ft. (10.5 m.)
Weight empty
  1,475 lb. (670 kg.)

PASSENGERS/CARGO
1 pilot and 1 passenger

MAX. CRUISING SPEED
115 m.p.h. (185 km.p.h.)

CEILING
14,000 ft. (4,300 m.) fully
  loaded

RANGE
445 miles (720 km.)

ENGINE
One 150 h.p. Blackburn
  Major II

CYGNET
1936

82

This side-by-side two seat training and club aircraft was an attempt at a more robust and easily made machine than those of longeron, plywood and fabric construction. It was the first British light aeroplane to have both metal skinned wings and fuselage. The first version had a single fin and rudder and a tailwheel landing gear. The Cygnet illustrated had twin-fins and a tricycle landing gear making it easier to fly.

BUILT BY
The Aichi Watch and Electric
   Machinery Co, Nagoya
DIMENSIONS
Length
   34.75 ft. (10.57 m.)
Wing span
   47.7 ft. (14.53 m.)
Weight empty
   5,770 lb. (2,617 kg.)
PASSENGERS/CARGO
1 pilot and 1 gunner
ARMAMENT
Two machine-guns
Max bomb load:
   550 lb. (250 kg.)
MAX. SPEED
270 m.p.h. (430 km.p.h.)
RANGE
840 miles (1,350 km.)
ENGINE
One 1, 045 h.p. Mitsubishi
   Kinsei

**VAL
1936**

A two-seat carrier based dive-bomber, the Aichi Type 99D3A—code named "Val"—spearheaded the Japanese surprise attack on Pearl Harbour in December, 1941. It caused immense damage and all but knocked out the American Pacific battle-fleet. Like other Japanese warplanes, however, it was lightly constructed and with its slow speed was easy prey for Allied fighters.

This general reconnaissance flying-boat, widely used by American and British forces during the Second World War, was ideally suitable for rescue from the sea, because of its endurance of over 17 hours, and high cruising speed. Successes with the R.A.F. included the spotting of the *Bismarck* after it had eluded Allied ships, and the sinking of the last U-boat of the war by Coastal Command—the 196th.

CATALINA 1936

BUILT BY
Consolidated Aircraft Corp, San Diego
DIMENSIONS
Length
65.1 ft. (19.83 m.)
Wing span
104 ft. (31.7 m.)
Weight empty
14,240 lb. (6,465 kg.)
PASSENGERS/CARGO
Crew 8 to 9
ARMAMENT
Four machine-guns
Max. bomb load:
2,000 lb. (900 kg.)
CRUISING SPEED
180 m.p.h. (290 km.p.h.)
CEILING
24,000 ft. (7,300 m.) fully loaded
RANGE
4,000 miles (6,400 km.)
ENGINES
2 × 1,200 h.p. Pratt & Whitney Twin-Wasp

GLADIATOR
1937

BUILT BY
Gloster Aircraft Co,
　Gloucester
DIMENSIONS
Length
　27.4 ft. (835 m.)
Wing span
　32.25 ft. (9.8 m.)
Weight empty
　3,450 lb. (1,565 kg.)
PASSENGERS/CARGO
1 pilot
ARMAMENT
Four fixed machine-guns,
　firing forward
MAX. SPEED
253 m.p.h. (407 km.p.h.)
CEILING
33,000 ft. (10,000 m.)
ENDURANCE
2 hours
ENGINE
One 840 h.p. Bristol
　Mercury IX

This, the last biplane fighter to serve with the R.A.F., represented the ultimate in its class; the efforts to reduce drag included the landing gear, the wheels of which not only enclosed the brakes but the leg shock-absorber as well. Gladiators operated during the Norwegian campaign in 1940, and in the heroic defence of Malta in June, 1940.

BUILT BY
Lockheed Aircraft
   Corporation, Burbank
DIMENSIONS
Length
   44.2 ft. (13.4 m.)
Wing span
   65.6 ft. (19.9 m.)
Weight empty
   10,300 lb. (4,670 kg.)
PASSENGERS/CARGO
2 crew; 12 passengers
MAX. CRUISING SPEED
230 m.p.h. (370 km.p.h.)
CEILING
24,000 ft. (7,300 m.) fully
   loaded
RANGE
2,125 miles (3,400 km.)
ENGINES
2 × 760 h.p. Wright Cyclone

LOCKHEED 14
1937

86

Based on the Electra, this was sometimes referred to as the ''Super Electra''. Basically similar to the earlier aircraft, it had more powerful engines and a more spacious fuselage. To help it operate from small airfields Fowler flaps were fitted to the wings. During the Second World War a military version, known as the ''Hudson'', was used by the R.A.F. for reconnaissance duties.

BUILT BY
Hawker Aircraft, Kingston
DIMENSIONS
Length
    31.4 ft. (9.55 m.)
Wing span
    40 ft. (12.2 m.)
Weight empty
    4,670 lb. (2,118 kg.)
PASSENGERS/CARGO
1 pilot
ARMAMENT
Mk. 1: 8 × 0.303 in.
    machine-guns
Mk. 2: 4 × 20 mm. cannon
Bomb load:
    1,000 lb. (450 kg.)
MAX. SPEED
316 m.p.h. (508 km.p.h.)
RANGE
460 miles (740 km.)
ENGINE
One 1,030 h.p. Rolls-Royce
    Merlin

HURRICANE
1937

87

During the Battle of Britain, the Hurricane shot down more enemy than all other aircraft and ground defences combined. It added to its laurels in Malta, the Western Desert and Burma, tending to ''hold the fort'', until enough Spitfires arrived to facilitate offensives. It proved effective for ground attack and troop support: it could carry bombs, rockets—or two deadly 40 mm. cannon for knocking out tanks.

Considered the most beautiful piston-engined aircraft ever built, this was designed as a transatlantic mailplane, had an exceptionally long range and high cruising speed, made possible because drag was reduced to the minimum. Five Albatross, equipped to carry passengers, were used as airliners on the busy Croydon to Paris, Brussels and Zurich routes. The two transatlantic machines were used by the R.A.F. during the Second World War on a shuttle service to Iceland.

ALBATROSS 1937

BUILT BY
The de Havilland Aircraft Co.,
  Hatfield
DIMENSIONS
Length
  71.5 ft. (21.8 m.)
Wing span
  105 ft. (32 m.)
Weight empty
  21,230 lb. (9,630 kg.)
PASSENGERS/CARGO
4 crew; 22 passengers
MAX. CRUISING SPEED
210 m.p.h. (340 km.p.h.)
CEILING
17,900 ft. (5,500 m.)
RANGE
3,300 miles (5,300 km.)
ENGINES
4 × 525 h.p. D.H. Gipsy
  Twelve

STRATOLINER
1938

BUILT BY
Boeing Airplane Co, Seattle
DIMENSIONS
Length
    74 ft. (22.6 m.)
Wing span
    107 ft. (32.6 m.)
Weight empty
    27,650 lb. (12,550 kg.)
PASSENGERS/CARGO
4 or 5 crew; 33 passengers
MAX. CRUISING SPEED
215 m.p.h. (344 km.p.h.)
CEILING 23,300 ft. (7,110
    km.) fully loaded
RANGE
1,635 miles (2,616 km.) at
    50 % maximum power
ENGINES
4 × 1,200 h.p. Pratt &
    Whitney Wasp

Based on the Flying Fortress bomber and efficiently and ruggedly constructed, the Boeing 307 was the first airliner with a pressurised cabin, helping to make it one of the most comfortable of airliners when it went into service in 1940, carrying passengers on long-haul American routes. The "night-plane" version had berths for 16 passengers and reclining chairs for a further 9.

BUILT BY
Sir W. G. Armstrong
  Whitworth Aircraft,
  Coventry

DIMENSIONS
Length
  114 ft. (34.7 m.)
Wing span
  123 ft. (37.5 m.)
Weight empty
  32,920 lb. (14,930 kg.)

PASSENGERS/CARGO
Up to 40 passengers

MAX. CRUISING SPEED
170 m.p.h. (270 km.p.h.)

CEILING
18,000 ft. (5,500 m.) fully
  loaded

Range
800 miles (1,300 km.)

ENGINES
4 × 850 h.p. Armstrong
  Siddeley Tiger

ENSIGN
1938
90

This four-engined, high-wing aircraft was the biggest liner used by the British airline Imperial Airways before the Second World War. Normally 27 passengers were accommodated in three cabins, but aircraft on short-haul European routes could carry up to 40. The aircraft was underpowered, therefore unsuccessful; to overcome this, four 950 h.p. Wright Cyclones were fitted. Ensigns were used for official duties during the war; several were lost.

BUILT BY
Supermarine, Southampton
DIMENSIONS
Length
29.9 ft. (9.1 m.)
Wing span
36.9 ft. (11.2 m.)
Weight empty
4,810 lb. (2,180 kg.)
PASSENGERS/CARGO
1 pilot
ARMAMENT
8 × 0.303 in. machine-guns
MAX. SPEED
355 m.p.h. (570 km.p.h.)
CEILING
34,000 ft. (10,360 m.)
fully loaded
RANGE
575 miles (920 km.)
ENGINE
One 1,030 h.p. Rolls-Royce
Merlin

The Spitfire, with its distinctive pointed wings, became a household word during the Battle of Britain. It soon proved itself better than the enemy and, with the Hurricane, saved Britain and the world. After the Battle, the improved versions developed demonstrated their superb qualities on every front. The last of the 20,000 Spitfires built, powered by Rolls-Royce Griffon engines, had a top speed of 460 m.p.h. (740 km.p.h.)

**SPITFIRE
1938**

In spite of being tail-heavy, owing to the two extra fins added because of trouble with the prototype, difficult to taxi because of the sponson-wings, and generally underpowered, this flying-boat was outstandingly successful. Three were obtained in 1941 by BOAC when Europe was at war and Britain was all but isolated from her friends and allies. The resulting regular, useful and morale-boosting flights to New York continued for six safe and efficient years.

BOEING 314
1938

BUILT BY
Boeing Airplane Co, Seattle
DIMENSIONS
Length
    106 ft. (32.3 m.)
Wing span
    152 ft. (46.3 m.)
Weight empty
    48,000 lb (21,772 kg.)
PASSENGERS/CARGO
11 crew; 44 passengers
MAX. CRUISING SPEED
188 m.p.h. (302 km.p.h.)
CEILING
24,000 ft. (7,300 m.) fully
    loaded
RANGE
3,685 miles (5,930 km.)
ENGINES
4 × 1,600 h.p. Wright
    Cyclone

FOCKE-WULF
F.W.190
1939

BUILT BY
Focke-Wulf Flugzeugbau,
 Bremen
DIMENSIONS
Length
 29 ft. (8.84 m.)
Wing span
 34.6 ft. (10.54 m.)
Weight empty
 7,000 lb. (3,175 kg.)
PASSENGERS/CARGO
1 pilot
ARMAMENT
2 × 7.9mm machine-guns
4 × 20 mm. cannon
MAX. SPEED
408 m.p.h. (655 km.p.h.)
CEILING
37,400 ft. (11,400 m.)
 fully loaded
RANGE
950 miles (1,500 km.)
ENGINE
One 1,700 h.p. BMW 801D

Of fine form, the radial engine being exceptionally neatly cowled, this was the best German single-seat fighter of the Second World War. The F.W.190G, a ground attack version, could carry a bomb load of 4,000 lb. (1,800 kg.) under the wings. Over 20,000 were built.

BUILT BY
Aeronautica D'Italia, Turin
DIMENSIONS
Length
 27.1 ft. (8.23 m.)
Wing span
 31.75 ft. (9.65 m.)
Weight empty
 3,763 lb. (1,707 kg.)
PASSENGERS/CARGO
1 pilot
ARMAMENT
Two machine-guns
MAX. SPEED
266 m.p.h. (428 km.p.h.)
CEILING
33,000 ft. (10,000 m.)
 fully loaded
RANGE
490 miles (790 km.)
ENGINE
One 840 h.p. Fiat A.74RC.38

FIAT CR.42 FALCO
1939
94

This aircraft has two dubious distinctions—it was the last single-seat biplane to be made by any of the Second World War combatants, and was the only Italian aircraft to take part in the Battle of Britain. It joined in a few attacks on Britain but, outclassed wherever it was used, it was relegated to night-fighting in Italy and fighter-bomber operations in North Africa.

BUILT BY
Short Brothers, Rochester
DIMENSIONS
Length
101.3 ft. (30.9 m.)
Wing span
134.3 ft. (41 m.)
Weight empty
37,700 lb. (17,100 kg.)
PASSENGERS/CARGO
7 crew; 40 passengers and
mail
CRUISING SPEED
180 m.p.h. (290 km.p.h.)
RANGE
3,200 miles (5,150 km.)
ENGINES
4 × 1,380 h.p. Bristol
Hercules IV

GOLDEN HIND
1939

95

The *Golden Hind* (officially the Short type 26) was one of three long-range flying-boats specially designed for non-stop services across the Atlantic. The war prevented this, and they were commandeered for long-range reconnaissance duties, for which gun turrets and depth charge housings were installed. In 1942 they were reconverted to carry up to 40 passengers and used on services to Africa.

This was the first of the big, four-engined heavy bombers to be used with such devastating effect against Germany in the Second World War. When design work started, no bomber capable of carrying such a huge load had been built, and in action the Stirling proved manoeuvrable and extremely tough. However, its low ceiling proved a handicap as did its bomb-bay, which, divided into sections, could only hold bombs of up to 4,000 lbs. (1,800 kg.).

STIRLING 1939

BUILT BY
Short Brothers, Rochester
DIMENSIONS
Length
    87.25 ft. (26.6 m.)
Wing span
    99 ft. (30 m.)
Weight empty
    43,200 lb. (19,590 kg.)
PASSENGERS/CARGO
Crew 7 or 8
ARMAMENT
Two machine-guns in each of
    the nose and dorsal
    turrets, four in tail turret
Max. bomb load:
    14,000 lb. (6,350 kg.)
MAX. SPEED
270 m.p.h. (434 km.p.h.)
RANGE
590 miles (950 km.)
ENGINES
4 × 1,650 h.p. Bristol
    Hercules

This, together with the Flying Fortress, was the main American heavy-bomber of the Second World War, and during the daylight bombing offensive against Germany dropped nearly half-a-million tons of bombs. Many of the 18,031 built served with the R.A.F. in anti-submarine reconnaissance, their long endurance enabling them to patrol the gap in mid-Atlantic where previously U-boats had been out of range of Allied land-based aircraft.

LIBERATOR
1939

BUILT BY
Consolidated Vultee Aircraft
Corporation, San Diego
DIMENSIONS
Length
67.1 ft. (20.42 m.)
Wing span
110 ft. (33.53 m.)
Weight empty
37,000 lb. (16,780 kg.)
PASSENGERS/CARGO
Crew: 12
ARMAMENT
10 × 0.50 in. machine-guns
Max. bomb load: 12,800 lb.
(5,800 kg.)
MAX. SPEED
270 m.p.h. (430 km.p.h.)
RANGE
990 miles (1,600 km.)
ENGINES
4 × 1,200 h.p. Pratt &
Whitney Twin Wasp

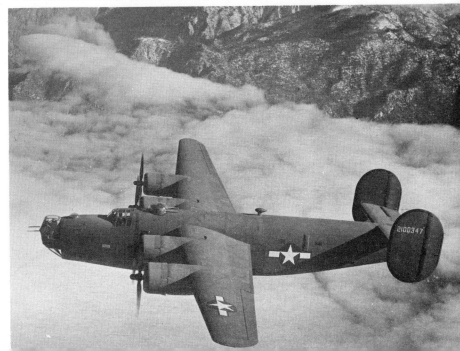

BUILT BY
United Aircraft Corporation,
  Stratford
DIMENSIONS
Length
  33.3 ft. (10.15 m.)
Wing span
  41 ft. (12.5 m.)
Weight empty
  8,980 lb. (4,070 kg.)
PASSENGERS/CARGO
1 pilot
ARMAMENT
6 × 0.50 in. machine-guns
2 × 1,000 lb. (450 kg.)
  bombs; or 8 × 5 in. (13
  cm.) rockets
MAX. SPEED
417 m.p.h. (670 km.p.h.)
RANGE
1,015 miles (1,630 km.)
ENGINE
One 2,000 h.p. Pratt &
  Whitney R2800

CORSAIR
1940

This single-seat fighter, achieved outstanding success during the Second World War, from both land bases and carriers, in spite of a high landing speed and a restricted view from the cockpit. Not the least important of its advantages was its sharply cranked wings, which helped to make its identification easy in the heat of battle. Later variants were armed with four 20mm. cannon and had more powerful engines.

BUILT BY
The de Havilland Aircraft
  Co, Hatfield
DIMENSIONS
Length
  41.5 ft. (12.65 m.)
Wing span
  54.2 ft. (16.5 m.)
Weight empty
  15,510 lb. (7,035 kg.)
PASSENGERS/CARGO
Crew: 2
ARMAMENT
Max. bomb load:
  4,000 lb. (1,800 kg.)
MAX. SPEED
408 m.p.h. (656 km.p.h.)
CEILING
37,000 ft. (11,000 m.) fully
  loaded
RANGE
1,370 miles (2,200 km.)
ENGINES
2 × 1,680 h.p. Rolls-Royce
  Merlin 72

Unarmed and of wooden construction the Mosquito is considered the most outstanding aircraft of the Second World War. The Air Ministry were initially uninterested in a bomber fast enough to do without guns, but when the prototype flew at 400 m.p.h. (640 km.p.h.) with fighter-like manoeuvrability, it was ordered into large-scale production. Later, the fighter and reconnaissance versions developed proved top of their classes.

MOSQUITO
1940

This single-seat fighter was the first American aircraft designed to satisfy R.A.F. requirements. Although the performance of the prototype proved far superior to contemporary American fighters, particularly at low altitudes, it was handicapped at height by the weak Allison engine. However, the installation of an American-built 1,680 h.p. Rolls-Royce Merlin made it one of the R.A.F.'s best fighters, and when fitted with drop tanks it had a remarkable operational range of over 1,700 miles (2,700 km.).

MUSTANG
1940

BUILT BY
North American Aviation,
  California
DIMENSIONS
Length
  32.1 ft. (9.76 m.)
Wing span
  37 ft. (11.28 m.)
Weight empty
  6,300 lb. (2,860 kg.)
PASSENGERS/CARGO
1 pilot
ARMAMENT
4 × 0.50 in. machine-guns,
  or 4 × 20 mm. cannon
MAX. SPEED
390 m.p.h. (630 km.p.h.)
CEILING
32,000 ft. (9,700 km.) fully
  loaded
RANGE
1,050 miles (1,690 km.)
ENGINE
One 1,150 h.p. Allison
  V-1710

LANCASTER
1941

This was the most successful bomber of the Second World War. 132 tons of bombs were dropped for each Lancaster lost, compared with 56 for each Halifax and 41 tons for each Stirling. Lancasters flew over 156,000 sorties and dropped 608,612 tons of bombs. Their exploits included the sinking of the *Tirpitz*, and they were the only bombers able to carry the R.A.F.'s big 12,000 lb. (5,500 kg.) and 22,000 lb. (9,980 kg.) "earthquake" bombs.

BUILT BY
A. V. Roe & Company,
  Manchester
DIMENSIONS
Length
  69.5 ft. (21.2 m.)
Wing span
  102 ft. (31.1 m.)
Weight empty
  36,900 lb. (16,740 kg.)
PASSENGERS/CARGO
Crew: 7
ARMAMENT
8 × 0.303 in. machine-guns
Max. bomb load: 22,000 lb.
  (9,980 kg.)
MAX. SPEED
287 m.p.h. (460 km.p.h.)
RANGE
1,660 miles (2,700 km.) with
  14,000 lb. (6,350 kg.) of
  bombs
ENGINES
4 × 1,460 h.p. Rolls-Royce
  Merlin

BUILT BY
Messerschmitt A.G.,
  Augsberg
DIMENSIONS
Length
  18.7 ft. (5.7 m.)
Wing span
  30.6 ft. (9.32 m.)
Weight empty
  4,200 lb. (1,900 kg.)
PASSENGERS/CARGO
1 pilot
ARMAMENT
2 × 30 mm. cannon, and
  24 RAM air-to-air rockets
MAX. SPEED
596 m.p.h. (960 km.p.h.)
CEILING
39,500 ft. (12,000 m.) fully
  loaded
ENDURANCE
8 minutes
ENGINE
One 3,750 lb. (1700 kg.)
  thrust Walten HWK 109

MESSERSCHMITT
Me.163
1941

This single-seat interceptor fighter was the most revolutionary of the Second World War. Its liquid-propellent rocket engine gave it a phenomenal rate of climb, a top speed much higher than that of any Allied aircraft, and its armament included air-to-air rockets. However its extremely short endurance limited its effectiveness. Also, it was difficult to fly and any residual propellents remaining in the tanks on landing often exploded, killing the pilot.

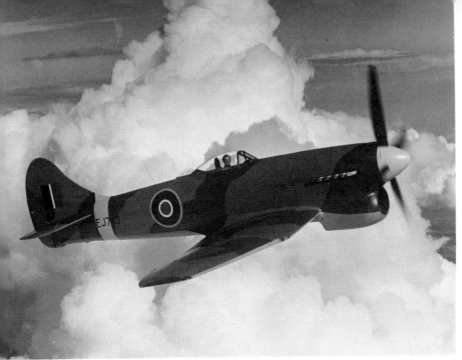

BUILT BY
Hawker Aircraft, Kingston
DIMENSIONS
Length
   33.7 ft. (10.27 m.)
Wing span
   41 ft. (12.5 m.)
Weight empty
   9,250 lb. (4,200 kg.)
PASSENGERS/CARGO
1 pilot
ARMAMENT
4 × 20 mm. cannon
2 + 1,000 lb. (450 kg.)
   bombs
2 × 60 lb. (30 kg.) rockets
MAX. SPEED
435 m.p.h. (700 km.p.h.)
CEILING
36,000 ft. (11,000 m.) fully
   loaded
RANGE
820 miles (1,320 km.)
ENGINE
One 2,180 h.p. Napier Sabre

Developed from the Typhoon, this single-seat interceptor embodied new semi-elliptical wings of thin section and a lengthened fuselage housing an additional fuel tank forward of the cockpit. The aircraft shot down 638 of the German V-1 flying bombs and later, armed with rockets or bombs, it proved an effective ground attack machine. In air battles it destroyed 20 of the Me. 262 jet-fighters.

TEMPEST
1942

During the Second World War, this Boeing B29 dropped 171,000 tons of bombs and incendiaries on the major Japanese towns, operating from surrounding islands captured specifically for the purpose. It is best remembered as the one used to drop atomic bombs on Hiroshima and Nagasaki, ending the war quickly without losing thousands of Allied troops.

SUPERFORTRESS 1942

BUILT BY
Boeing Airplane Co.
    Seattle
DIMENSIONS
Length
    99 ft. (30.2 m.)
Wing span
    141.25 ft. (43.1 m.)
Weight empty
    70,140 lb. (31,810 kg.)
PASSENGERS/CARGO
Crew 12
ARMAMENT
12 × 0.50 in. machine-guns
Max. bomb load:
    20,000 lb. (9,000 kg.)
MAX. SPEED
360 m.p.h. (580 km.p.h.)
RANGE
3,250 miles (5,230 km.) with
    10,000 lb. of bombs
ENGINES
4 × 2,200 h.p. Wright
    R.3350

MESSERSCHMITT Me.262 1942

The Me.262 single-seat fighter was the first turbojet aircraft to attack other aircraft—Britain's jet Meteor went into action first, but against V-1 flying bombs. Over 1,400 were built, but only about 200 used operationally. Allied bombing raids and Hitler's decision to try and turn all the Me. 262s into fighter-bombers prevented it from having any serious effect on the Allied bombing offensive.

BUILT BY
Messerschmitt A. G.,
  Augsberg
DIMENSIONS
Length
  34.8 ft. (10.6 m.)
Wing span
  41 ft. (12.5 m.)
Weight empty
  8,370 lb. (3,800 kg.)
PASSENGERS/CARGO
1 pilot
ARMAMENT
4 × 30 mm. cannon
24 RAM air-to-air rockets
MAX. SPEED
540 m.p.h. (870 km.p.h.)
CEILING
37,500 ft. (11,500
  m.) fully loaded
RANGE
650 miles (1,050 km.)
ENGINES
2 × 1,980 lb. (900 kg.)
  thrust Junkers Jumo 004B

DESIGNED BY
Alexander Sergievich
 Yakovlev, USSR
DIMENSIONS
Length
 27.9 ft. (8.5 m.)
Wing span
 32.75 ft. (9.96 m.)
Weight empty
 6,000 lb. (2,720 kg.)
PASSENGERS/CARGO
1 pilot
ARMAMENT
1 × 20 mm. cannon and
 1 × 12.7 mm. machine-
 gun
MAX. SPEED
360 m.p.h. (580 km.p.h.)
CEILING
36,000 ft. (11,000 m.)
 fully loaded
RANGE
880 miles (1,400 km.)
ENGINE
One 1,260 h.p Klimov
 VK-105PF

YAKOVLEV YAK-9D
1943

Nearly 30,000 of these single-seat fighters were made during the Second World War, and although unsophisticated, its rugged construction proved ideal for the arduous conditions under which it had to operate, (including those in the Battle of Stalingrad), and it was highly popular with its pilots. Some Yaks were fitted with either a 37 mm. or 75 mm. cannon for anti-tank use.

DIMENSIONS
Length
    41.3 ft. (12.6 m.)
Wing span
    43 ft. (13.1 m.)
Weight empty
    8,140 lb. (3,690 kg.)
PASSENGERS/CARGO
1 pilot
ARMAMENT
4 × 20 mm. cannon
MAX. SPEED
410 m.p.h. (660 km.p.h.)
CEILING
44,000 ft. (13,400 m.) fully
    loaded
RANGE
1,340 miles (2,150 km.)
ENGINES
2 × 1,700 lb. (770 kg) thrust
    Rolls-Royce Welland

This single-seat fighter, (the only Allied jet aircraft used in action in the Second
World War), joined the piston-engined Tempests and Spitfires against the V-1 flying bombs
in 1944. On later aircraft the Welland engines were replaced by 2000 lb. (900 kg.) and
3500 lb. (1,600 kg.) thrust Derwents, giving top speeds of 490 m.p.h. (790 km. p.h.) and
550 m.p.h. (885 km.p.h.) respectively, and were the backbone of Fighter Command until
superseded by Hunters in 1955.

**METEOR
1943**

The Nakajima Type 4, Ki 84 Hayate—code named "Frank"—was a single-seat fighter ordered to counter the high performance U.S. fighters and bombers. It was more robust than other Japanese fighters, and was able to out-turn contemporary versions of the Thunderbolt and Mustang. Later versions, armed with 20 mm. or 30 mm. cannon in place of the normal fuselage-mounted machine-guns, were even more formidable.

FRANK
1943

BUILT BY
Nakajima Aircraft Co,
   Tokyo
DIMENSIONS
Length
   32.5 ft. (9.9m.)
Wing span
      36.8 ft. (11.21 m.)
Weight empty
   5,860 lb. (2,660 kg.)
PASSENGERS/CARGO
1 pilot
ARMAMENT
2 × 20 mm. cannon, and 2 ×
   12.7 mm. machine-guns
Plus 1,100 lb. (500 kg.) of
   bombs
MAX. SPEED
390 m.p.h. (630 km.p.h.)
RANGE
1,080 miles (1,740 km.)
ENGINES
One 1,900 h.p.
   Nakajima Ha 45/11

This was the R.A.F.'s second jet aircraft. The unusual twin-boom layout was adopted to keep the length of the jet tailpipe as short as possible, to minimise power losses. It was easy to fly, had pleasant handling qualities, and was ideal for introducing pilots to jet flying. From the Vampire there developed the more advanced Venom series of fighters which in turn led to the near-supersonic Sea Vixen.

VAMPIRE
1943

BUILT BY
The de Havilland Aircraft Co,
  Hatfield
DIMENSIONS
Length
  30.75 ft. (9.35 m.)
Wing span
  40 ft. (12.2 m.)
Weight empty
  6,370 lb. (2,890 kg.)
PASSENGERS/CARGO
1 pilot
ARMAMENT
4 × 20 mm. cannon
MAX. SPEED
540 m.p.h. (870 km.p.h.)
CEILING
40,000 ft. (12,000 m.) fully
  loaded
RANGE
730 miles (1,200 km.)
ENGINE
One 3,100 lb. (1,400 kg.)
  thrust D.H. Goblin

BUILT BY

The de Havilland Aircraft Co,
  Hatfield

DIMENSIONS

Length
  36.7 ft. (11.18 m.)

Wing span
  45 ft. (13.72 m.)

Weight empty
  12,820 lb. (5,815 kg.)

PASSENGERS/CARGO

1 pilot

ARMAMENT

4 × 20 mm. cannon
Plus rockets or 2,000 lb.
  (900 kg.) bomb load

MAX. SPEED

470 m.p.h. (760 km.p.h.)

CEILING

35,000 ft. (11,000 km.)
  fully loaded

RANGE

3,000 miles (4,800 km.)

HORNET
1944
110

This single-seat fighter was intended for an island-hopping campaign against Japan in the South Pacific, but the war ended before any of them reached the Far East. They carried piston-engined performance to the near-ultimate, were the fastest such aircraft to serve in the R.A.F., and the last to remain in service in first-line squadrons. A two-seat version was produced for the Royal Navy.

BUILT BY
Ernst Heinkel Flugzeugwerke,
  Rostock
DIMENSIONS
Length
  29.8 ft. (9.08 m.)
Wing span
  23.7 ft. (7.2 m.)
Weight
  5,480 lb. (2,485 kg.)
PASSENGERS/CARGO
1 pilot
ARMAMENT
2×30 mm. cannon
MAX. SPEED
520 m.p.h. (840 km.p.h.)
CEILING
39,500 ft. (12,000 m.) fully
  loaded
RANGE
410 miles (660 km.)
ENGINE
One 1,760 lb. (800 kg.)
  thrust BMW 003A turbojet

Officially the Heinkel He.162, popularly the ''Volksjager'' or ''People's Fighter'', the Salamander was designed and built in the fantastically short period of sixty-nine days. It was a high-performance fighter, produced with the minimum of difficult-to-get materials by semi-skilled mechanics. Owing to the disruptive Allied bombing only 116 were built, although many more were in an advanced state of completion in underground factories at the end of the war.

SALAMANDER
1944

This was a courageous attempt to build a small aeroplane that could carry usefully-bulky loads over short ranges. The tubby fuselage was thus functional rather than pretty, and could carry a small car. It was not outstandingly successful, partly because of the early state of air freight business and because of an inadequate performance on one engine.

AEROVAN
1944

BUILT BY
Miles Aircraft, Reading
DIMENSIONS
Length
36 ft. (10.97 m.)
Wing span
50 ft. (15.24 m.)
Weight empty
3,410 lb. (1,546 kg.)
PASSENGERS/CARGO
2 crew, plus 8 passengers or
2,000 lb. (900 kg.) freight
MAX. CRUISING SPEED
110 m.p.h. (180 km.p.h.)
CEILING
13,000 ft. (4,000 km.) fully
loaded
RANGE
450 miles (720 km.)
ENGINES
2 × 155 h.p. Blackburn Cirrus
Major

TUDOR
1945

BUILT BY
A. V. Roe & Co,
    Manchester
DIMENSIONS
Length
    79.5 ft. (24.23 m.)
Wing span
    120 ft. (36.58 m.)
Weight empty
    47,960 lb. (21,750 kg.)
PASSENGERS/CARGO
12–42 passengers
CRUISING SPEED
210 m.p.h. (340 km.p.h.)
CEILING
26,000 ft. (8,000 km.) fully
    loaded
RANGE
3,600 miles (5,800 km.)
ENGINES
4 × 1,770 h.p. Rolls-Royce
    Merlin

The Tudor set a pattern which has bedevilled British aircraft in this class ever since. Conceived as the first new post-war civil airliner, the prototype revealed snags requiring expensive—and time-wasting—modifications. After two years of non-stop alterations, BOAC called for a further 343 modifications. Two Tudors were lost without trace, casting doubt on their airworthiness, although later aircraft, with independent airlines, proved reliable and useful.

BUILT BY
Bachem Werke
DIMENSIONS
Length
  21.25 ft. (6.47 m.)
Wing span
  13 ft. (3.96 m.)
Weight
  4,800 lb. (2,180 kg.)
PASSENGERS/CARGO
1 pilot
ARMAMENT
33 × 50 mm. RAM unguided
  rockets
MAX. SPEED
560 m.p.h. (900 km.p.h.)
ENDURANCE
2 minutes
ENGINES
One 3,750 lb. (1,700 kg.)
  thrust Walten HWK 109–
  509 rocket motor and
  four jettisonable
  Schmidding rockets of
  1,100 lb. (500 kg.) thrust

NATTER
1945
114

This was a rocket-propelled, single-seat fighter, to protect "pin-point" targets. It was to take off nearly vertically, climb under ground control until within 1 mile (1.6 km.) of the enemy, when the pilot would take over and fire the missiles. He and the rear fuselage, containing the rocket motor, would then descend by parachute. Several were tested successfully, but ten set up for defence were destroyed by their German ground crews before use owing to the approach of Allied ground forces.

BUILT BY
The de Havilland Aircraft Co,
 Hatfield
DIMENSIONS
Length
 39.25 ft. (11.95 m.)
Wing span
 57 ft. (17.37 m.)
Weight empty
 6,325 lb. (2,870 kg.)
PASSENGERS/CARGO
2 crew and up to 11
 passengers
CRUISING SPEED
200 m.p.h. (320 km.p.h.)
CEILING
21,700 ft. (6,600 m.) fully
 loaded
RANGE
500 miles (800 km.)
ENGINES
2 × 330 h.p. D.H. Gipsy
 Queen 70

Replacing the pre-war biplane Rapide, this was de Havilland's first peace-time project. It proved ideal as a small feederliner and as a company business aircraft; one such aircraft flew 30,000 miles (50,000 km.) and carried 300 passengers in four months. Well over 500 were sold, many for export. Later versions had more powerful engines and a redesigned cockpit canopy, as in the photograph.

DOVE
1945

115

This was the first British post-war aeroplane to enter airline service and was the mainstay of BEA until the Viscount turboprop airliner. For speed and cheapness parts of the Wellington bomber were used, particularly in the wings. The fuselage, of stressed-skin construction, was new and very strong. On one flight in 1950 a bomb almost blew off the tail, yet the pilot's skill plus the strength of the fuselage ensured a safe landing.

VIKING 1945

BUILT BY
Vickers-Armstrong,
   Weybridge
DIMENSIONS
Length
   65.1 ft (19.8 m.)
Wing span
   89.25 ft. (27.2 m.)
Weight empty
   22,910 lb. (10,400 kg.)
PASSENGERS/CARGO
2 crew and 21 passengers
CRUISING SPEED
210 m.p.h. (340 km.p.h.)
CEILING
22,000 ft. (6,700 km.) fully
   loaded
RANGE
1,875 miles (3,000 km.)
ENGINES
2 × 1,690 h.p. Bristol
   Hercules

This was a luxury four-seat private and club aircraft which made a promising start, but failed to hold its place against fierce American competition. It was a popular entrant for air races but few major successes were gained. About 150 were built in the first year of production, most being bought by owners overseas. A wide variety of engines were fitted and a large number of modifications made to the aircraft.

GEMINI
1945

BUILT BY
Miles Aircraft, Reading
DIMENSIONS
Length
    22.25 ft. (6.77 m.)
Wing span
    36.1 ft. (11 m.)
Weight empty
    1,900 lb. (860 kg.)
PASSENGERS/CARGO
1 pilot; 3 passengers
CRUISING SPEED
125 m.p.h. (200 km.p.h.)
CEILING
13,500 ft. (4,100 m.) fully
    loaded
RANGE
820 miles (1,320 km.)
ENGINES
2 × 100 h.p. Blackburn Cirrus
    Minor

BUILT BY

The de Havilland Aircraft
of Canada Ltd, Toronto

DIMENSIONS

Length
  25.7 ft. (7.82 m.)
Wing span
  34.3 ft. (10.45 m.)
Weight empty
  1,425 lb. (646 kg.)

PASSENGERS/CARGO

1 pilot and 1 passenger

MAX. SPEED

140 m.p.h . (225 km.p.h.)

CEILING

16,700 ft. (5,000 m.) fully
  loaded

RANGE

280 miles (450 km.)

ENGINE

One 145 h.p. D.H. Gipsy
  Major 10

CHIPMUNK
1946
118

Intended to replace the immensely popular Tiger Moth, this was a military aircraft designed and built by the Canadian de Havilland factory. Adopted as the standard R.A.F. trainer, it was also built at the de Havilland factories at Hatfield and Chester. Although pleasant to fly and ideal for aerobatics, it was relatively expensive and was thus not widely used for civilian purposes.

BUILT BY
Hawker Aircraft, Kingston
DIMENSIONS
Length
   39.7 ft. (12.1 m.)
Wing span
   39 ft. (11.89 m.)
PASSENGERS/CARGO
1 pilot
ARMAMENT
4 × 20 mm. cannon
MAX. SPEED
620 m.p.h. (1,000 km.p.h.)
CEILING
50,000 ft. (15,000 m.) fully
   loaded
RANGE
1,000 miles (1,600 km.)
ENGINE
One 5,000 lb. (2,270 kg.)
   thrust Rolls-Royce Nene
   turbojet

This was the first jet combat aircraft produced by Hawker, famous for their brilliant fighters for 25 years. A novel feature was the tail jet-pipe which divided and exhausted in the wing roots. This made it exceptionally manoeuvrable and allowed a large internal fuel capacity giving the fighter a relatively long range. In addition to the fuselage-mounted guns, bombs and rockets could be carried under the wings.

SEA HAWK
1947

120

Using swept-wings, based on German research, and an RD-45 turbojet engine based on the Rolls-Royce Nene, the Mig-15 was a good fighter in its day. Very manoeuvrable with a high rate of climb, it served as the standard jet-fighter in the Red Air Force, and was exported in large numbers to neighbouring Communist countries. From it has developed a succession of formidable, high-performance Mig fighters.

MIG 15
1947

DESIGNED BY
Artem Mikoyan and Mikhail
  Gurevich
DIMENSIONS
Length
  36.3 ft. (11.06 m.)
Wing span
  33.1 ft. (10.1 m.)
Weight empty
  8,320 lb. (3,770 kg.)
PASSENGERS/CARGO
1 pilot
ARMAMENT
1 × 37 mm. cannon and
  2 × 23 mm. cannon.
MAX. CRUISING SPEED
670 m.p.h. (1,080 km.p.h.)
CEILING
51,000 ft. (15,500 m.)
  fully loaded
ENGINE
One 6,000 lb. (2,700 kg.)
  thrust Russian-built Nene
  turbojet

SABRE
1947

More than 6,000 of these single-seat fighters were built. Of these over 2,000 were F-86Ds, with a nose radome for all-weather fighter duties and rocket armament instead of machine-guns. The later F-86L had extended wings, leading-edge slats and electronic equipment integrating them into the sophisticated American SAGE air defence system. Sabres were extensively used by foreign air forces.

BUILT BY
North American Aviation
DIMENSIONS
Length
37.5 ft. (11.33 m.)
Wing span
37.1 ft. (11.3 m.)
Weight
18,000 lb. (8,160 kg.)
PASSENGERS/CARGO
1 pilot
ARMAMENT
6 × 0.50 in. machine-guns
MAX. SPEED
650 m.p.h. (1,050 km.p.h.)
CEILING
45,000 ft. (14,000 m.)
fully loaded
RANGE
900 miles (1,450 km.)
ENGINE
One 5,200 lb. (2,360 kg.)
thrust General Electric
J-47 turbojet

BUILT BY
Vickers-Armstrong,
  Weybridge
DIMENSIONS
Length
  81.2 ft. (24.75 m.)
Wing span
  94 ft. (28.65 m.)
Weight, loaded
  56,000 lb. (2,540 kg.)
PASSENGERS/CARGO
3 to 4 crew; 40 to 48
  passengers
MAX. CRUISING SPEED
300 m.p.h. (480 km.p.h.)
CEILING
28,500 ft. (8,700 km.) fully
  loaded
RANGE
750 miles (1,200 km.) with
  13,000 lb. payload
ENGINES
4 × 1,530 h.p. Rolls-Royce
  Dart turboprop

VISCOUNT
1948
122

The Viscount 700 went into service with BEA in 1950, when its smooth ride and high speed attracted passengers away from the airlines operating older, vibration-prone, piston-engined planes. As the world's first turboprop transport it was particularly trouble-free. The 800 series had more powerful engines and a longer fuselage seating up to 75 passengers. This was the most successful British transport ever; 438 were built.

BUILT BY
Handley Page, Cricklewood
DIMENSIONS
Length
    96.8 ft. (29.5 m.)
Wing span
    113 ft. (34.43 m.)
Weight empty
    55,350 lb. (25,100 kg.)
PASSENGERS/CARGO
5 crew; 40 passengers
CRUISING SPEED
275 m.p.h. (440 km.p.h.)
CEILING
24,500 ft. (7,500 m.) fully
    loaded
RANGE
2,000 miles (3,200 km.) with
    14,000 lb (6,350 kg.)
    payload
ENGINES
4 × 2,100 h.p. Bristol
    Hercules 763

HERMES
1948

Designed during the Second World War for future peace-time routes, the Hermes was dogged by misfortune and delays. The initial prototype of 1945 stalled and crashed, and it was not until September, 1948 that the definitive passenger version flew. Even that did not enter service with BOAC until 1950, by which time the type was obsolescent. It was replaced by Canadian C-4s in 1952.

123

This development of the Lincoln bomber was a long-range maritime reconnaissance and anti-submarine aircraft, the first British four-engined aeroplane to fly with contra-props. The main production version was the M.R. Mk.3, with nose-wheel landing gear, increased fuel capacity, wing tip tanks, improved cockpit canopy and large retractable radome located under the fuselage aft of the bomb-bay. It was superseded by the Hawker Siddeley Nimrod.

SHACKLETON
1949

BUILT BY
A.V. Roe and Co,
   Manchester
DIMENSIONS
Length
   92.5 ft. (28.19 m.)
Wing span
   119.9 ft. (36.52 m.)
Weight empty
   57,800 lb. (26,218 kg.)
PASSENGERS/CARGO
Crew: 10
ARMAMENT
2 × 20 mm. cannon in nose,
   plus large load of anti-
   submarine weapons, air/
   sea rescue lifeboat, etc.
MAX. SPEED
302 m.p.h. (486 km.p.h.)
RANGE
4,215 miles (6,780 km.) at
   200 m.p.h. (320 km.p.h.)
ENGINES
4 × 2,455 h.p. Rolls-Royce
   Griffon 57

This, the world's first jet airliner, is assured of a permanent place in aviation history, just failing to become also one of the great aeroplanes of all time. One of the best looking aircraft ever designed, it cut journey times by half, but crashed twice over the Mediterranean. Redesign work resulted in the Comet 4, with Rolls-Royce Avons giving more than twice the power, payload and longer ranges. However in the time lost America had gained her present lead in big jet transport.

COMET
1949

BUILT BY
The de Havilland Aircraft Co,
   Hatfield
DIMENSIONS
Length
   93 ft. (28.35 m.)
Wing span
   115 ft. (35 m.)
Weight
   105,000 lb. (45,540 kg.)
PASSENGERS/CARGO
36—44 first class passengers
MAX. CRUISING SPEED
490 m.p.h. (790 km.p.h.)
CRUISING ALTITUDE
35,000 ft. (11,000 km.)
   fully loaded
RANGE
1,750 miles (2,800 km.)
ENGINES
4 × 4,450 lb. (2,020 kg.)
thrust D.H. Ghost 50
turbojet

BUILT BY
English Electric Co, Preston
DIMENSIONS
Length
  65.5 ft. (19.96 m.)
Wing span
  64 ft. (19.5 m.)
Weight empty
  23,170 lb. (10.510 kg.)
PASSENGERS/CARGO
Crew 2 or 3
ARMAMENT
Max. bomb load: 6,000 lb.
  (2,700 kg.)
MAX. SPEED
580 m.p.h. (930 km.p.h.)
CEILING
48,000 ft. (14,600 m.) fully
  loaded
RANGE
3,000 miles (4,800 km.)
ENGINES
2 × 7,400 lb. (3,360 kg.)
  thrust Rolls-Royce Avon
  109 turbojet

CANBERRA
1949
126

This was the first jet bomber produced in Britain and to serve with the R.A.F., and was unarmed, relying on speed and altitude to escape interception. It was exceptionally manoeuvrable; this led to an intruder version with four 20 mm. or 30 mm. cannon and four 1,000 lb. (450 kg.) bombs in an optional ventral pack. A similar version was built under licence in America—the first British combat aircraft so built since the First World War.

BUILT BY
Boeing Airplane Co, Seattle
DIMENSIONS
Length
110.3 ft. (33.62 m.)
Wing span
141.25 ft. (43 m.)
Weight empty
83,500 lb. (37,870 kg.)
PASSENGERS/CARGO
7 crew and up to 100
passengers
MAX. CRUISING SPEED
340 m.p.h. (550 km.p.h.)
CEILING
32,000 ft. (10,000 m.) fully
loaded
RANGE
4,600 miles (7,400 km.)
ENGINES
4 × 3,500 h.p. Pratt &
Whitney R-4360

This was a passenger development of the B-29 Superfortress bomber of the Second World War, the conversion being effected by placing a large luxurious passenger cabin on top of the bomber fuselage! It was popular, and most successful in bridging the gap between the end of the war and the time when new airliners became available. A military version, used by the U.S. Air Force, was known as the "Stratofreighter".

STRATOCRUISER 1949

128

A light, tactical bomber, the IL-28 was first shown publicly in 1950, although it had entered service with the Red Air Force in 1949 and the prototype flew well before then. Intended to fulfil the same function as Britain's Canberra, it did not rely on altitude and speed to escape interception, hence its formidable defensive armament. Several thousand were built and the bomber was also supplied to foreign air forces.

ILYUSHIN IL-28
1950

DESIGNED BY
Sergei Vladimirovich Ilyushin, USSR

DIMENSIONS
Length
62 ft. (18.9 m.)
Wing span
68 ft. (20.75 m.)
Weight empty
26,000 lb. (11,795 kg.)

PASSENGERS/CARGO
Crew 4

ARMAMENT
2 × 23 mm. cannon in nose and 2 × 23 mm. cannon in tail. Max. bomb load: 4,400 lb. (2,000 kg.)

MAX. SPEED
580 m.p.h. (935 km.p.h.)

RANGE
1,500 miles (2,400 km.) with max. bomb load

ENGINES
4 × 5,955 lb. (2,700 kg.) thrust VK-1 turbojet

SUPER CONSTELLATION 1950

This development of the Model 649 "standard" Constellation actually started in 1939 when Lockheed were attempting to meet the requirements of TWA. Modified into a military transport during the Second World War, the design was reconverted into an airliner which came at the right time to equip airlines starved of new aircraft by the war. The distinctive upswept tail and downswept nose of the fuselage were adopted in the interests of aerodynamic efficiency.

BUILT BY
Lockheed Aircraft
Corporation, Burbank

DIMENSIONS
Length
113.6 ft. (34.62 m.)
Wing span
123 ft. (37.47 m.)
Weight, max
137,500 lb. (62,370 kg.)

PASSENGERS/CARGO
6 crew and 65–89
passengers

MAX. CRUISING SPEED
310 m.p.h. (500 km.p.h.)

RANGE
4,800 miles (7,700 km.) with
18,000 lb. payload

ENGINES
4 × 3,400 h.p. Wright
R-3350 turbo-compound
piston-engines.

BUILT BY
Hawker Aircraft Co., Kingston

DIMENSIONS
Length
45.9 ft. (13.99 m.)
Wing span
33.67 ft. (10.26 m.)
Weight empty
13,270 lb. (6,020 kg.)

PASSENGERS/CARGO
1 pilot

ARMAMENTS
4 × 30 mm. cannon, plus
bombs and rockets

MAX. SPEED
715 m.p.h. (1,150 km.p.h.)

CEILING
55,000 ft. (16,750 m.) fully
loaded

RANGE
1,840 miles (2,965 km.)

ENGINE
One 10,000 lb. (4,500 kg.)
thrust Rolls-Royce Avon
turbojet.

HUNTER
1951

130

This single-seat fighter superseded the Meteor Mk. 8 as the standard day interceptor in Fighter Command. Variants included a two-seat trainer version, a Royal Naval version with an arrester hook, and aircraft powered by a Bristol Siddeley Sapphire turbojet. Many pilots preferred the Hunter to later and ostensibly more advanced types, and second-hand machines are eagerly sought by the many foreign air forces in which it still serves.

BUILT BY
A.V. Roe and Co,
   Manchester
DIMENSIONS
Length
   99.9 ft. (30.45 m.)
Wing span
   111 ft. (33.83 m.)
PASSENGERS/CARGO
Crew: 5
ARMAMENT
Max. bomb load: 21,000 lb.
   (9,525 kg.), which can
   include nuclear weapons
   and the Blue Steel stand-
   off bomb
MAX. CRUISING SPEED
Mach 0.94
MAX. CRUISING HEIGHT
55,000 ft. (16,750 m.) fully
   loaded
ENGINES
4 × 20,000 lb. (9,072 kg.)
   thrust Rolls-Royce Bristol
   Olympus turbojet.

This was the first jet-bomber in the world to employ the delta-wing plan form. The proto-type was powered by four Rolls-Royce Avon turbojets and had a ''simple'' leading edge. Later aircraft had redesigned wings with the distinctive compound sweepback, and an extended and bulged tail cone containing anti-missile radar-jamming electronic equipment. Three squadrons are equipped with the Blue Steel air-to-surface rocket-powered, nuclear-armed bomb.

VULCAN B. Mk. 2
1952

131

The outstanding feature of the Victor is its crescent-shaped wing. On this the degree of sweep reduces progressively to the tip, and provides good characteristics at high altitude for long range, combined with good control at low speeds for taking off and landing. Like its Vulcan sister-ships, Victors have been camouflaged for low-level under-the-radar-screen strikes, and some have served as in-flight refuellers and reconnaissance planes.

VICTOR B. Mk. 2
1952

BUILT BY
Handley Page, Cricklewood
DIMENSIONS
Length
114.9 ft. (35 m.)
Wing span
120 ft. (36.6 m.)
PASSENGERS/CARGO
Crew: 5
ARMAMENT
Max. bomb load: 35,000 lb.
(15,875 kg.)
NORMAL CRUISING
SPEED
Mach 0.92
MAX. CRUISING HEIGHT
55,000 ft. (16,750 m.) fully
loaded
COMBAT RADIUS
2,300 miles (3,700 km.) at
high level.
ENGINES
4 × 20,600 lb. (9,344 kg.)
thrust Rolls-Royce Conway
Turbofan

B. 52 STRATOFORTRESS 1952

This huge long-range strategic bomber formed the backbone of America's ''deterrent'' force, serving with Strategic Air Command until superseded by the silo-sunk Minuteman and submarine-borne Polaris intercontinental ballistic missiles. Stratofortresses were used extensively in Vietnam, where they were used to bomb suspected concentrations of enemy troops and supply routes.

BUILT BY
Boeing Airplane Co, Seattle
DIMENSIONS (B. 52-H)
Length
156 ft. (47.55 m.)
Wing span
185 ft. (56.42 m.)
Weight max
488,000 lb. (221,350 kg.)
PASSENGERS/CARGO
Crew: 6
ARMAMENTS
One 20 mm. cannon in tail, rocket launchers, stand-off bombs and missiles
MAX. SPEED
650 m.p.h. (1,040 km.p.h.)
CEILING
60,000 ft. (18,300 m.) fully loaded
MAX. RANGE
12,500 miles (20,120 km.)
ENGINES
8 × 17,000 lb. (7,720 kg.) thrust Pratt & Whitney TF33 turbofan

BUILT BY
Bristol Aeroplane Co, Bristol
DIMENSIONS (300)
Length
   124.25 ft. (37.89 m.)
Wing span
   142.25 ft. (43.38 m.)
Weight empty
   93,100 lb. (42.230 kg.)
PASSENGERS/CARGO
Crew: 9 and up to 133
   passengers
**LONG–RANGE CRUISING
   SPEED**
355 m.p.h. (570 km.p.h.)
RANGE
4,268 miles (6,869 km.) with
   max. payload of 28,000 lb.
   (12,700 kg.)
ENGINES
4 × 4,400 h.p. Bristol
   Siddeley Proteus 705
   turboprop engine

BRITANNIA
1952
134

This, the ultimate in propeller-driven airliners, was intended to bridge the gap between the old piston-engined craft and the jet-liners. Unfortunately, although ordered in 1948, continued changes of mind and extended development trials, kept the aircraft out of service until 1957. This was too late for its undoubted qualities to be fully exploited, and it did not attract the anticipated world-wide airline orders. R.A.F. Transport Command operates 20 Britannias, with large freight doors, as all-purpose transports.

BUILT BY
British Aircraft Corporation,
 Preston
DIMENSIONS (F. Mk. 6)
Length
 55.25 ft. (16.84 m.)
Wing span
 34.8 ft. (10.61 m.)
PASSENGERS/CARGO
1 pilot
ARMAMENT
2 × 30 mm. cannon and two
 Redtop missiles, plus 48 ×
 2 in. (5 cm.) unguided
 rockets or additional
 cannon
MAX. SPEED
Mach 2.0
ENGINES
2 × 16,360 lb. (7,420 kg.)
 thrust Rolls-Royce Avon
 turbojet

LIGHTNING
1952

135

Unusual features of this, Britain's first truly supersonic fighter, are the placing of the two engines one above the other in the fuselage and the location of the ailerons on the tips of the sharply swept wings. The fully-developed production version has an extended outer leading-edge, to reduce drag, and a new ventral fuel pack with more than twice the original capacity. An arrester hook is fitted for emergency use. Export versions are heavily armed.

136

This, the first jet transport to be built in the United States is the airliner which, together with the Douglas DC-8, brought large scale jet travel to the world. Over 800 were produced for the U.S. Air Force, providing valuable experience for the famous family of civil versions which have followed. Many improvements have been embodied in the Model 320B.

BOEING 707
1954

BUILT BY
Boeing Airplane Co, Seattle
DIMENSIONS
Length
152.9 ft. (46.61 m.)
Wing span
145.75 ft. (44.42 m.)
Weight empty
138,385 lb. (62,771 kg.)
PASSENGERS
Max. accommodation: 189
economy class passengers
CEILING
42,000 ft. (12,800 m.) fully
loaded
RANGE
6,160 miles (9,915 km.) no
reserves
ENGINES
4 × 18,000 lb. (8,165 kg.)
thrust Pratt & Whitney
JTD-3 turbofan

The Tu.104 was Russia's first jet airliner, and is a passenger-carrying development of the Tu.16 bomber, embodying a bigger fuselage. The Tu.104 accommodated 50 passengers, the Tu.104A and the Tu.104B, with a slightly longer fuselage, seats 100. Although it went into service as long ago as 1956, it is still widely used in Russia and her neighbouring countries.

TUPOLEV Tu. 104a 1955

DESIGNED BY
Andrei Nikolaevich Tupolev
DIMENSIONS
Length
    126.3 ft. (38.50 m.)
Wing span
    113.3 ft. (34.54 m.)
Max. T.O. Weight
    166,450 lb. (75,500 kg.)
PASSENGERS
100 passengers
MAX. CRUISING SPEED
560 m.p.h. (900 km.p.h.)
CRUISING HEIGHT
39,000 ft. (12,000 m.) fully
    loaded
RANGE
2,610 miles (4,200 km.) with
    17,640 lb. (8,000 kg.)
    payload
ENGINES
2 × 19,180 lb. (8,700 kg.)
    thrust Mikolin AM-3M
    turbojet

BUILT BY
Svenska Aeroplan AB,
  Linkoping
DIMENSIONS
Length
  51.9 ft. (15.80 m.)
Wing span
  30.9 ft. (9.40 m.)
T.O. Weight
  19,800 lb. (9,000 kg.)
PASSENGERS/CARGO
1 pilot
ARMAMENT
2 × 30 mm. cannon and two
  or four Sidewinder missiles,
  plus bombs or rockets
MAX. SPEED
1,320 m.p.h. (2,125 km.p.h.)
CEILING
60,000 ft. (18,000 m.) fully
  loaded
ENGINE
One 17,635 lb. (8,000 kg.)
  thrust Rolls-Royce Avon
  300 turbojet

DRAKEN
1955
138

This is a highly efficient supersonic short-range interceptor and strike fighter. Of interest is its double-delta wing platform, developed exclusively in Sweden, which provides aerodynamic efficiency and a relatively large internal space for fuel, armament and other equipment. Electronic equipment includes a collision-course system to assist in the interception of bombers under all weather conditions.

BUILT BY
Sud Aviation, Toulouse
DIMENSIONS
Length
    105 ft. (32.01 m.)
Wing span
    112.5 ft. (34.30 m.)
Weight empty
    57,935 lb. (26,280 kg.)
PASSENGERS/CARGO
Max. accommodation: 99
    tourist-class passengers
MAX. CRUISING SPEED
525 m.p.h. (845 km.p.h.)
RANGE
1,430 miles (2,300 km.) with
    max. payload
ENGINES
2 × 12,600 lb. (5,725 kg.)
    thrust Rolls-Royce Avon
    turbojet

CARAVELLE
1955

This twin-jet, short-to-medium range airliner is the only aircraft of its type to be developed in the West outside the United States and Britain. Of interest is the finely streamlined nose, based on that of Britain's Comet jetliner. The latest model is the Series IIR, for mixed passenger-freight, with American Pratt and Whitney turbofan engines, and a large freight door in the left side of the fuselage.

This is a very successful medium-sized, short-to-medium range airliner which has been in production for several years in both the United States and Holland. The engine and much of the internal equipment is supplied by Britain, including the landing gear shock absorber struts, wheels and brakes. Series 500 aircraft have a lengthened fuselage for 56 passengers. The military version can carry 45 paratroops or 24 stretchers and 7 attendants.

FRIENDSHIP
1955

BUILT BY
Fokker, Amsterdam
DIMENSIONS
Length
    77.1 ft. (23.5 m.)
Wing span
    95.2 ft. (29.0 m.)
Weight empty
    23,200 lb. (10,525 kg.)
PASSENGERS
52 passengers
MAX. CRUISING SPEED
295 m.p.h. (474 km.p.h.)
CEILING
28,500 ft. (8,700 m.) fully
    loaded
RANGE
1,285 miles (2,070 km.)
ENGINES
2 × 2,050 h.p. Rolls-Royce
    Dart turboprop

This is a single-seat interceptor, tailored closely to enable it to climb quickly and shoot accurately in the most efficient manner. It can climb at about 50,000 ft. (15,000 m.) per minute at sea level, and can accelerate to Mach 2 in three minutes. It is a formidable weapon in the hands of a skilled pilot but is a tricky craft for novices and has a high accident rate during training.

STARFIGHTER 1956

BUILT BY
Lockheed Aircraft
  Corporation, Burbank
DIMENSIONS
Length
  51.25 ft. (15.62 m.)
Wing span
  21.9 ft. (6.68 m.)
Weight empty
  14,082 lb. (6,387 kg.)
PASSENGERS/CARGO
1 pilot
ARMAMENT
Two or four missiles, plus
  rockets or bombs
MAX. SPEED
1,450 m.p.h. (2,330 km.p.h.)
CEILING
58,000 ft. (17,680 m.) fully
  loaded
RANGE
745 miles (1,200 km.)
ENGINE
One 15,800 lb. (7,165 kg.)
  thrust General Electric J79
  turbojet

BUILT BY
Hadinka Works, Moscow
DIMENSIONS
Length
  117.75 ft. (35.9 m.)
Wing span
  122.7 ft. (37.4 m.)
Weight empty
  77,160 lb. (35,000 kg.)
PASSENGERS
Max. capacity: 122
  passengers
MAX. CRUISING SPEED
420 m.p.h. (675 km.p.h.)
RANGE
2,300 miles (3,700 km.) with
  max. payload
ENGINES
4 × 4,250 h.p. Ivchenko
  AI-20 turboprop

ILYUSHIN IL-18
1957
142

One of the better Soviet transports built since the end of the war, this aircraft has an impressive performance and, being rugged and reliable, has been supplied to many foreign airlines. It is estimated that more than 500 have been built, and the aircraft has been used to help develop the Polosa automatic landing system, which meets western standards.

BUILT BY
Lockheed Aircraft
  Corporation, California
DIMENSIONS
Length
  104.5 ft. (31.81 m.)
Wing span
  99 ft. (30.18 m.)
Weight empty
  57,300 lb. (25,990 kg.)
PASSENGERS
High density version: 99
  passengers
MAX. CRUISING SPEED
405 m.p.h. (650 km.p.h.)
CEILING
28,400 ft. (8,655 m.)
RANGE
2,770 miles (4,458 km.) with
  18,000 lb. (8,165 kg.)
  payload
ENGINES
4 × 4,050 h.p. Allison 501-
  D15 turboprop

Nearly 200 of this medium-range airliner, designed specifically for airline routes in the United States, were produced and the type served a useful purpose pending the development of short-range jet airliners. It is in widescale service (in Europe with the Dutch airline KLM) and large numbers of "Orion", an anti-submarine version, have been produced.

ELECTRA
1957

This is a twin-engined, two-seat, carrier-borne tactical strike and all-weather interceptor aircraft, not beautiful, but unsurpassed as a combat warplane. It can carry a formidable array of underwing bombs and other stores, making it a deadly ground attack machine. Those supplied to the Royal Navy, powered by Rolls-Royce Spey turbofan engines, have an even better performance than their American counterparts.

PHANTOM II
1958

BUILT BY
McDonnell Co, St. Louis
DIMENSIONS
Length
    58.25 ft. (17.76 m.)
Wing span
    38.5 ft. (11.70 m.)
PASSENGERS/CARGO
1 pilot, 1 navigator
ARMAMENT
Six Sparrow or four Sparrow
    and four Sidewinder
    missiles, or up to 16,000
    lb. (7,250 kg.) of
    underwing stores
MAX. SPEED
Mach 2+
CEILING
71,000 ft. (21,600 m.)
RANGE
1,000 miles (1,600 km.)
ENGINES
2 × 16,500 lb. (7,485 Kg.)
    thrust General Electric J-79
    turbojet

DOUGLAS DC-8
1959

This was the first jet transport to be produced by Douglas. It has been progressively developed, the latest version, the Super 63, having a fuselage lengthened by 36.7 ft. (11.18 m.), an extended wing span, improved engine pods and redesigned engine pylons to reduce drag. The maximum range of this version is 7,700 miles (12,400 km.).

BUILT BY
The Douglas Aircraft Co,
  Santa Monica
DIMENSIONS (Super 63)
Length
  187.4 ft. (57.12 m.)
Wing span
  148.4 ft. (45.23 m.)
Weight empty
  153,749 lb. (69,739 kg.)
PASSENGERS/CARGO
Max. capacity: 251
  passengers
MAX. CRUISING SPEED
600 m.p.h. (965 km.p.h.)
RANGE
4,600 miles (7,400 km.) with
  max. payload
ENGINES
4 × 19,000 lb. (8,618 kg.)
  Pratt & Whitney JT3D
  engine

BUILT BY
Hawker Siddeley Aviation,
Chester
DIMENSIONS
Length
47.4 ft. (14.45 m.)
Wing span
47.0 ft. (14.33 m.)
Weight empty
11,900 lb. (5,400 kg.)
PASSENGERS/CARGO
2 crew and up to 8
passengers
MAX. CRUISING SPEED
505 m.p.h. (813 km.p.h.)
CEILING
39,000 ft. (11,900 m.) fully
loaded
RANGE
1,940 miles (3,120 km.) with
max. payload
ENGINES
2 × 3,360 lb. (1,525 kg.)
thrust Bristol Siddeley
Viper turbojet

HAWKER SIDDELEY
HS 125
1962

This is a twin-jet executive aircraft used by large or widely scattered companies who require rapid transport without booking scheduled airlines. It is extremely comfortable and the roomy cabin enables passengers to walk to their seats without stooping. Nearly 200 had been sold in America by the end of 1969 and a navigational trainer version, the "Dominie", is in service with the R.A.F.

BUILT BY
Hawker Siddeley Aviation,
Hatfield

DIMENSIONS
Length
114.75 ft. (34.98 m.)
Wing span
98 ft. (29.87 m.)
Weight empty
73,550 lb. (33,361 kg.)

PASSENGERS
BEA aircraft: 94 tourist class
passengers

MAX. CRUISING SPEED
605 m.p.h. (972 km.p.h.)

RANGE
2,450 miles (3,950 km.)
with passenger capacity
payload

ENGINES
3 × 11,930 lb. (5,410 kg.)
Rolls-Royce Spey turbofan

The Trident was designed to meet BEA requirements for a short-haul, 600 m.p.h. (965 km.p.h.) airliner for use during 1965–75. The 2E is the long-range version with more fuel, operating non-stop between London and the Middle East. Under development is the 3B, with a lengthened fuselage seating up to 146 passengers and, in addition to the three Spey engines, a Rolls-Royce RB 162 booster engine in the tail.

TRIDENT 2E
1962

This is the standard long-range jet airliner in service with Aeroflot, the Russian national airline. It first entered service, after a long series of snags, in 1967, when it was introduced on the long Moscow-Montreal route. During the development period one of the many changes made was the embodiment of the distinctive saw-tooth wing leading-edge.

ILYUSHIN IL-62
1963

DESIGNED BY
Sergei Vladimirovich Ilyushin, USSR
DIMENSIONS
Length
174.25 ft. (53.12 m.)
Wing span
142 ft. (43.30 m.)
Weight empty
149,475 lb. (67,800 kg.)
PASSENGERS/CARGO
Max. capacity: 186 passengers
MAX. CRUISING SPEED
560 m.p.h. (900 km.p.h.)
RANGE
5,715 miles (9,200 km.) with 22,050 lb. (10,000 kg.) payload
ENGINES
4 × 23,150 lb. (10,500 kg.) Kuznetsov NK84 turbofan

The early days of this short-to-medium range airliner were marred by crashes during landing, but, shortening journey times and being more comfortable than the propeller-driven aircraft it replaced, the liner attracted many passengers. Also it made money for the airlines with the result that today hundreds are in service all over the world. The latest version, the "Series 200", (to which the accompanying details apply), has a lengthened fuselage.

BOEING 727
1963

BUILT BY
The Boeing Co, Seattle
DIMENSIONS
Length
153.2 ft. (46.69 m.)
Wing span
108 ft. (32.92 m)
Weight empty
89,300 lb. (40,500 kg.)
PASSENGERS
Max. capacity: 189
passengers
MAX. CRUISING SPEED
592 m.p.h. (953 km.p.h.)
CEILING
35.200 ft. (10,730 km.) fully
loaded
RANGE
2,300 miles (3,700 km.) with
25,000 lb. (11,340 kg.)
payload
ENGINES
3 × 14,000 lb. (6,350 kg.)
Pratt & Whitney JT8D
turbofan

BUILT BY
British Aircraft Corporation,
    Weybridge
DIMENSIONS
Length
    171.7 ft. (52.32 m.)
Wing span
    146.2 ft. (44.55 m.)
Weight empty
    154,552 lb. (70,104 kg.)
PASSENGERS
Max. capacity: 174 economy
    class passengers
MAX. CRUISING SPEED
581 m.p.h. (935 km.p.h.)
RANGE
4,720 miles (7,600 km.)
ENGINES
4 × 22,500 lb. (10,205 kg.)
    Rolls-Royce Conway
    turbofan

SUPER VC-10
1964
150

Acclaimed as the most comfortable and quietest airliner, the VC-10, designed to operate from small and difficult airfields in Africa, has excellent take-off characteristics and a relatively low landing speed. It is thus not quite as efficient as other aircraft over long ranges. The Super VC-10, with a longer fuselage, an improved wing leading-edge and more powerful engines, carries more passengers for a small increase in take-off distance.

DESIGNED BY
Artem Mikoyan and Mikhail
  Gurevich, USSR
DIMENSIONS
Length
  69 ft. (21.0 m.)
Wing span
  40 ft. (12.2 m.)
PASSENGERS/CARGO
1 pilot
ARMAMENT
Not known
MAX. SPEED
1,800 m.p.h.
  (2,900 km.p.h.) approx.
CEILING
100,000 ft. (30,000 km.)
  approx.
RANGE
1,000 miles (1,600 km.)
  estimated
ENGINES
2 × 24,000 lb. (11,000 kg.)
  thrust turbofan

This is the latest of the successful combat aircraft by Russian designers Mikoyan and Gurevich. Few details of this twin-jet, all-weather fighter are available, but an early version, the ''E-266'', gained some impressive records, including speed and payload-to-height records in 1967. Perhaps the most noteworthy aspect of this undoubtedly formidable aircraft is its general similarity to Britain's cancelled TSR-2.

**MIG 23
1965**

151

152

It was to be expected that Douglas would not let the Boeing 727 and the BAC One-Eleven fulfil all the need for short-range jet airliners. Their contender, the DC-9, was ordered in such large numbers that Douglas had to merge with McDonnell to build them all. In service the DC-9 is proving reliable and economical. The most popular variant is the "Series 30", to which the accompanying details apply.

DOUGLAS DC-9
1965

BUILT BY
The Douglas Aircraft Co,
 Santa Monica

DIMENSIONS
Length
 119.25 ft. (36.37 m.)
Wing span
 93.4 ft. (28.47 m.)
Weight empty
 52,935 lb. (24,010 kg.)

PASSENGERS
Series 20: 90 passengers
Series 30:115 passengers
Series 40:125 passengers

MAX. CRUISING SPEED
565 m.p.h. (909 km.p.h.)

RANGE
1,725 miles (2,775 km.)
 with 50 passengers

ENGINES
2 × 14,000 lb. (6,350 kg.)
 thrust Pratt & Whitney JT8D
 turbofan

Developed from the Kestrel, this is the world's first VTOL fixed-wing combat aircraft: it can take off and land vertically, by means of four nozzles, rotated to deflect the engine exhaust downwards for take-off and when airborne, rearwards, when the Harrier flies conventionally. This enables it to operate from small, easily concealed, patches of flat land, close to areas of fighting.

HARRIER
1966

BUILT BY
Hawker Siddeley Aviation,
 Kingston
DIMENSIONS
Length
 46 ft. (14.02 m.)
Wing span
 25 ft. (7.62 m.)
PASSENGERS/CARGO
1 pilot
ARMAMENT
Max. weapon load: 4,000 lb.
 (1,815 kg.). Typical load:
 108 × 68 mm. rockets,
 2 × 30 mm. cannon and
 one 500 lb (225 kg.) bomb
MAX. SPEED
Mach 0.87 estimated
FERRY RANGE
2,300 miles (3,700 km.) with
 underwing tanks
ENGINE
One 19,000 lb. (8,620 kg.)
 thrust Rolls-Royce Pegasus
 vectored thrust turbofan

DESIGNED BY
Alexander Sergievich Yakolev,
  USSR
DIMENSIONS
Length
  66.25 ft. (20.19 m.)
Wing span
  82 ft. (25.0 m.)
T.O. Weight
  30,200 lb. (13,700 kg.)
PASSENGERS
Max. capacity: 31 passengers
CRUISING SPEED
342 m.p.h. (550 km.p.h.)
RANGE
1,025 miles (1,650 km.)
ENGINES
3 × 3,300 lb. (1,500 kg.)
  thrust Ivchenko A1-25
  turbofan

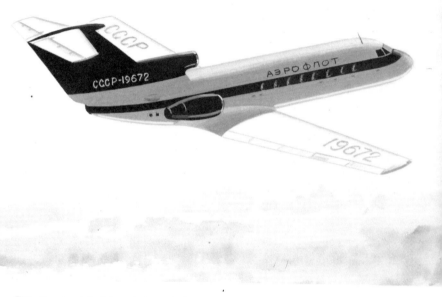

YAK 40
1966
154

This Russian jet airliner is intended to replace Aeroflot's Russian-built Dakotas, and its most interesting feature is that although very small it is powered by three tail-mounted turbofan engines. To increase its usefulness and to enable it to be used from minor airports, it is designed to operate from grass airfields. When it is used as a freighter, the passenger seats fold back against the sides of the cabin.

The Viggen, ("Thunderbolt") is an extremely advanced multi-mission combat aircraft under development for the Royal Swedish Air Force. The unusual wing planform comprises a foreplane combined with a main delta wing, and embodies flap-blowing to increase lift. This gives the aircraft good take-off and landing characteristics, so that it can operate from small landing strips. A number of variants are planned.

BUILT BY
SAAB Aktiebolag, Linköping
DIMENSIONS
Length
    53.5 ft. (16.30 m.)
Wing span
    34.75 ft. (10.60 m.)
T.O. Weight
    35,275 lb. (16,000 kg.)
PASSENGERS/CARGO
1 pilot
ARMAMENT
Air-to-surface missiles,
    bombs, guns and mines,
    air-to-air missiles for
    interception duties
MAX. SPEED
Mach 2+
ENGINE
One 26,450 lb. (12,000 kg.)
    Pratt & Whitney JT8D
    turbofan

VIGGEN
1967

Orders by American operators represented a major tribute to the technical excellence of this short-range airliner, designed as a jet successor to the Viscount. It has proved remarkably reliable and the cheapest airliner in the world to operate. The latest version, the Series 500, has a lengthened fuselage, increased wing span and more powerful engines. BEA refer to this model as the "Super One-Eleven".

SUPER ONE-ELEVEN 1967

BUILT BY
British Aircraft Corporation,
   Weybridge
DIMENSIONS
Length
   107 ft. (32.61 m.)
Wing span
   93.5 ft. (28.5 m.)
Weight empty
   53,995 lb. (24,490 kg.)
PASSENGERS/CARGO
Max. capacity: 99
   passengers
MAX. CRUISING SPEED
548 m.p.h. (882 km.p.h.)
MAX. CRUISING HEIGHT
35,000 ft. (10,670 m.)
RANGE
576 miles (927 km.) with
   typical payload and two
   hours' reserve fuel
ENGINES
2 × 11,970 lb. (5,434 kg.)
   thrust Rolls-Royce Spey
   turbofan

The "jumbo-jet" can accommodate 490 passengers, although initial versions have a basic layout for about 350 passengers in greater comfort. The wide and high passenger compartments set a fashion which is being followed by the Lockheed 10—11 and McDonnell Douglas DC-10 shorter range jumbo-jets. The distinctive flight deck was placed up high so that the whole nose can hinge upward to allow nose-in loading on the all-freighter version.

BOEING 747
1969

BUILT BY
The Boeing Co, Seattle
DIMENSIONS
Length
231.3 ft. (70.51 m.)
Wing span
195.7 ft. (59.64 m.)
Max. weight 710,000 lb.
(322,000 kg.)
PASSENGERS
Max. capacity: 490
passengers
MAX. CRUISING SPEED
640 m.p.h. (1,030 km.p.h.)
CEILING
45,000 ft. (13,000 m.) fully
loaded
RANGE
4,600 miles (7,400 km.)
with maximum payload.
ENGINES
4 × 43,500 lb. (19,730 kg.)
thrust Pratt & Whitney JT9D
turbofan

BUILT BY
Sud Aviation and the British
    Aircraft Corporation
DIMENSIONS
Length
    193 ft. (58.84 m.)
Wing span
    84 ft. (25.6 m.)
Weight empty
    159,500 lb. (72,350 kg.)
PASSENGERS
Max. capacity: 128
    passengers
MAX. CRUISING SPEED
1,450 m.p.h. (2,333 km.p.h.)
CEILING
65,000 ft. (19,800 m.) fully
    loaded
RANGE
4,250 miles (6,830 km.)
    with maximum payload
ENGINES
4 × 37,400 lb. (16,900 kg.)
    thrust Bristol Siddeley
    Olympus turbofan

CONCORDE
1969
158

Designed and built jointly by France and Britain, this is the world's second supersonic
airliner, after Russia's Tu. 144. The basic simplicity of the exterior tends to hide the
complexity of the internal systems and equipment. The fuel system, for example, is used
also to maintain the aircraft trim while it accelerates and slows down, and to cool the
cabin air. The Concorde is taking aviation to the limits of present day materials, processes
and techniques.

INDEX

159